Total Food
Sustainability of the Agri-Food Chain

Total Food
Sustainability of the Agri-Food Chain

Edited by

K. W. Waldron, G. K. Moates
Institute of Food Research, Norwich, UK

C. B. Faulds
Centro de Investigaciones Biológicas (CSIC), Madrid, Spain

RSC Publishing

The proceedings of an International Conference held at the Norwich BioSciences Conference Centre, Norwich, UK on 22-24 April 2009.

Special Publication No. 323

ISBN: 978-1-84755-750-6

A catalogue record for this book is available from the British Library

Published by The Royal Society of Chemistry,
Thomas Graham House, Science Park, Milton Road,
Cambridge CB4 0WF, UK

Registered Charity Number 207890

For further information see our web site at www.rsc.org

Preface

The "Total Food" series of biennial, international conferences was initiated in 2004 by the Royal Society of Chemistry Food Group and the Institute of Food Research, Norwich. The aim of Total Food is to debate global research and development relevant to exploiting the whole food crop rather than the limited proportion that is consumed at present. For example, many vegetables are subjected to a high degree of trimming during post-harvest processing, and many parts including the outer leaves and stems are often disposed of by landfill. Much of the 'wasted' material is potentially food grade and may also contain important ingredients for further exploitation. The volumes involved are considerable. In the EC, over 500,000 tonnes of onion waste are disposed of per year. In some cases, up to 65% of food-grade material may be discarded during processing, and the EC-funded "AWARENET" report (2004) indicated that up to 222 million tonnes of food chain waste from all food processing sectors are produced annually across the EU.

Since the Total Food series began, the issue of food security has become prominent. The increasing global population in conjunction with the use of crops for biofuel production mean that the more efficient exploitation of biomass will be required. The Total Food conferences are well placed to provide regular forums to highlight recent developments and to facilitate knowledge transfer between representatives of the agri-food (and increasingly non-food) industries, scientific research community, legal experts on food-related legislation and waste management, and consumer organisations.

We gratefully acknowledge financial sponsorship from the following organisations without whose help the 2009 conference would not have been possible: the International Agri-Technology Centre, the Institute of Food Research, the Food & Health Network, Norwich Research Park and Achor International.

Information and downloads:

Total Food 2009 Conference details, abstracts and presentations may be obtained from the website: http://www.ifr.ac.uk/totalfood2009/

Total Food 2004: Presentations, abstracts and the complete Proceedings Volume may be downloaded from the website http://www.totalfood2004.com/

K.W. Waldron
G.K. Moates
C.B. Faulds

Contents

Key Drivers

A SOCIO-ECONOMIC PERSPECTIVE ON CO-PRODUCT EXPLOITATION

B. Gremmen[1,2], P. van Haperen[2] and J. Lamerichs[2]

[1]Laboratory of Plant Breeding, Wageningen University, PO Box 386, Droevendaalsesteeg 1, 6708 PB, Wageningen. E-mail: bart.gremmen@wur.nl
[2]Centre for Methodical Ethics and Technology Assessment, Wageningen University, Hollandseweg 1, 6706 KN, Wageningen

1 INTRODUCTION

Each year very large quantities of food processing waste co-products are produced within the EU. There are four ways to dispose of these co-products: incineration, landfill, animal feed, and composting. However, there is a possible fifth way: the upgrading of co-products. These co-products contain potentially marketable components including nutrients and micronutrients, rheological agents, texturised residues, flavours and colorants. These could be utilised in pharmaceutical, cosmetic and nutraceutical high-value products, as well as in contributing to medium value food and feed ingredients. The economic recovery and up-grading of these components from co-products to provide value-added food-grade commodities not only promises considerable profits but is also vital to sustainable development and to address environmental concerns about waste disposal. Despite these great expectations, agro waste co-products are currently a substantial negative cost for the food industry.

Within the research framework of the EU the REPRO project has developed advanced and integrated (bio) processing methods to enable total transformation of food-derived plant-based co-products into higher value food and feed products.[1] The research has exploited two major food-processing waste co-products which have had relatively little research and development input, but which comprise major waste streams across the EC (and globally): Spent Grain (from the Brewing Industry) representing "hard" cereal residues, and Vegetable trimmings (from the vegetable processing industry) representing "soft" plant food wastes.

The underlying drivers have been to achieve the objective of REPRO whilst avoiding microbiological hazards, ensuring that the co-product processing stream has inbuilt traceability and that the final products comply with existing food regulations and are acceptable to the consumer (environmentally and socioeconomically).

The aim of this paper is to report on the socioeconomic research of REPRO. A team of social scientists has performed an evaluation of the preferences of different stakeholders, and an assessment of the economic feasibility and efficiency of the processes and products. Specific emphasis was paid to differences in concerns and representations of issues between different stakeholders.

Four different social research activities were carried out over a period of three years. In a pilot study a literature review and interviews were used to lay down general concepts for further use. Two Delphi studies, one among the REPRO partners and one among representatives from industry, were performed to determine which elements were more critical in evaluation. Focus group discussions were held to test the relevance of those elements in the eyes of consumers. Finally, a cost/benefit analysis has been performed.

2 PILOT STUDY

2.1. Literature review

The aim of the pilot study was to identify central themes that make up consumers' views of potential risks and benefits attached to new food processing technologies and the introduction of novel (modified or functional) food products. A literature review was made that deals with concepts around waste and dirt, in order to investigate what possible elements can be found here that could form contributing success or failure factors in relation to the nature of the raw material (waste) used to develop novel foods.

The starting point for the literature review does not lie in the specifics of the REPRO project in particular but in the broader issue that REPRO addresses: how do consumers perceive sustainability initiatives such as upgrading agro food-waste.[2] It will also address to what extent consumers judge the application of novel production technologies, and how they come to value sustainable end products in relation to other existing products. Key areas for the study were Novel foods, novel production techniques, risk communication and risk disputes, and sustainability.

There seems to be a partial overlap between novel foods and novel production methods. The absence of tangible products with desirable benefits is seen to have a negative influence. Consumers are presented to perceive food production practices, products and product features as well as possible risks involved, in terms of a trade off between a broad array of competing norms and values. Public risk perception in literature starts to be related not only to food safety issues, but also to lifestyle hazards. The difference between lay and expert perceptions of risk remains valid, although theoretical models about the effect and how to handle the difference vary. Therefore it is suggested to deal with consumer risk perceptions in their own right, rather than in relation to the expert view. Trust plays a greater role than the exact knowledge consumers have in forming their opinion. Consequently, also the effect of providing information in that sense was related to trusting the source, rather than evaluating the quality of information. Extra information reinforced already existing opinions, rather than changing the course thereof. Literature about benefits of functional foods[2] indicates that the opinions about benefits might be construed locally and in context. The image of the basic product plays a role, as well as the cultural association people make with the targeted marketing group. Sustainability is conceptually ambiguous, covering a range of practices and interests.

In the literature several definitions of waste can be found. One possible definition of waste in the context of REPRO is: "Any substance or object the holder discards, intends to discard or is required to discard" is WASTE under the Waste Framework Directive (European Directive 2006/12/EC). Once a substance or object has become waste, it will remain waste until it has been fully recovered and no longer poses a potential threat to the environment or to human health. The above definition handles two dimensions of waste that could be related to cultural values. 1) The moment the wish arrives to discard of something, and 2) The moment discarded matter is recovered or not regarded as waste any longer.

Jean Beaudrillard[3] relates waste to affluence in the sense that wasting necessarily has become a expression of status. They who can afford to waste most, win prestige. Beaudrillard sees wasting as the inherent logic of an economic system that has folded back into itself: destruction as the modern engine for economic growth. Creating waste or discarding matter in this light becomes an expression of affluence and from here on it can be expected that consumers possess a need to show affluence by becoming more particular of what they eat and what they throw away.

Mary Douglas[4] brings up the perspective of classification of matter. Dirt would be "matter out of place". Ordering our world is partly done on the basis of classification of matter and behaviour in physical space. It thus not only determines how matter is regarded, but also where it should be physically placed. Matter out of place is disturbing this order and therefore is culturally contaminating. The place where it is stored or dumped or retrieved from stigmatises. Food becomes dirty when it is found in the wrong place, such as a waste container. Unclean, untidy, dirty, these are denominations created by matter not being in the place where it belongs. This could imply that a bad carrot can still be used as it is taken from a case of carrots, but not any longer when it is taken from a landfill. Waste as a category comes into play whenever residue in the food industry is too problematic to keep within the limits of food safety regulations. Isolation, contamination, etc. all have to be kept under check. Regulatory restrictions and environmentalism are the expression that waste has become just that which doesn't belong in either nature or (physical) society and therefore also has become something that is per definition out of place till some method or strategy has been found to reposition or re-classify it in such a way that it falls under either nature or culture again. This can either be a technical solution to rework the waste, or a shift in classification.

The idea that (reprocessed) co-products become culturally acceptable once they are assigned a proper place in our day to day life is underpinned by already existing products, which do not carry negative associations, for example bran and germ (from the milling of whole grains into refined grains), pectin (recovered from the remains of processed fruit), collagen and gelatin (from the boiled skin and other parts of slaughtered livestock), and whey (from cheese manufacturing).

The theory of Mary Douglas could be very relevant for consumer research, as it implies that when co-products receive a decent classification and a decent place in the production process, they do not contaminate anymore, therefore would not be considered waste any longer. However, in line with Douglas's theory, there is a hierarchy of waste discernable from waste that can easily be handled to waste that cannot be touched, nearing taboo. The disgust waste provokes, is directly related to taboo. Waste in relation to taboo carries the connotation of excretion. In this way it relates to the extent to which it is processed, decomposed, digested, etc.

Viewed in respect to co-products of the processing industry it could thus be expected that there is the following hierarchy: left over produce after selection (small apples, crooked carrots etc), being of insufficient quality; left over produce after cleaning, cutting (cabbage cuts, cauliflower leaves etc), or left over produce after peeling (shells of pulses, shafts etc); being leftover, residue, etc.; left over produce after processing (BSG, etc), waste, spoils, garbage, etc.

The theory of Beaudrillard implies that even if technology is developed to reprocess by-products, it might have little appeal to consumers who, from a desire to show affluence, need a certain amount of "wasting". It also hints at a prediction that reprocessing will not reduce the absolute amount of waste, as waste production is needed to feed a growing economy. Reprocessing would only create an opportunity to waste more somewhere else.

2.2 Interviews with retailers

A series of nine interviews was held with retailers from speciality food shops in the Netherlands (Wageningen and Eindhoven) and the United Kingdom (Glasgow) to present their views on the REPRO project. First, the respondents were asked to read one of two texts on the REPRO approach: one text being written in a very popular language and containing a number of highly predictable puns, and another text representing more factual information in a neutral way. The aim was to see whether the way information is brought influences the respondents' opinion.

What did the respondents think about the REPRO approach? The idea of reusing waste already exists but is nevertheless found interesting by retailers. The REPRO approach gives a new solution to the food waste problem, which is defined as 'mountains and mountains' of waste. Only one retailer thought it would be a trivial enterprise to reuse vegetable trimmings as the basis for new food products.

The name 'waste' or 'by-product' should not be used when presenting food products to consumers. The name suggests that these food products come from bins or are scraped from the floor. 'By-product' is a term to technical for the 'people on the street', many people will not understand what it means, and it can easily be associated with "some kind of chemical manipulation". Moreover connotations are made with BSE.

What were the retailers' concerns? An important concern was about the technology used to reprocess waste: you cannot see what has been done to the product and what chemicals are being used during the process. The use of chemicals is a worrisome aspect. Also the length of the process (if the waste product undergoes many steps it will not be beneficial), the amount of waste produced by reprocessing waste, and the amount of time before waste is used / reprocessed were mentioned.

According to the retailers they will be less negative about reusing waste in food products than consumers. Retailers know the processes, know that there are many checks, consumers do not. The retailers were worried about the extent to which big companies will turn the REPRO idea into a big profit making enterprise, and about the amount of money charged for products containing waste. The products are price-prohibitive: when compared to the 'initial' product, 'waste' products are too expensive. Consumers do not want to be fooled: when they buy an expensive, well-known brand they do not want to find a 'waste' product.

Asked about food safety/ hygiene, only one retailer had concerns on food safety / hygiene. Most retailers think that in a serious production process this will be dealt with sufficiently. Retailers had positive as well as negative opinions on the appeal of health enhancing ingredients on consumers.

The retailers were interested in selling products containing waste in their specialty food shop or supermarket. Food waste products combine both an interesting solution to decrease the amount of food waste as well as providing some health benefits. Moreover the novelty of these products appealed to the retailers. Retailers think there is only a limited market for food products containing waste. These kinds of products attract a specific market, consumers who already are aware and / or are against mass production. Buying these products is an ideological choice. On the other hand sales also depend on how a business presents itself and their products. If well-presented and when the attention of consumers is drawn it will also appeal to the 'average' consumer. Most consumers however have a strong view on waste as being something that belongs in the bin and has to stay there.

Retailers think that product appearance, taste, easiness of transportation e.g. is product small enough to take with you on a picnic or during cycling, brand name, and fair trade will influence consumers' buying behaviour. Consumers will choose healthy 'waste' juices over healthy 'waste' snacks: "a juice appeals more to customers than some exotic crisps". The reason is that soups and juices are known, but exotic / unknown snacks need persuasion to be sold, consumers do not know the taste. In order to sell the snacks, samples must be supplied to consumers. Additional information on the REPRO project is needed. For example about the way in which the REPRO idea will be achieved logistically: how it will operate on a larger scale. Also, more information is needed about the process that waste is going through to become a new food product.

3 TWO DELPHI STUDIES

3.1 Delphi study among REPRO partners

On the basis of the literature review and the interviews a Delphi study was developed among the REPRO partners. The format and formula of Delphi studies has developed ever since the instrument was first employed as a forecasting method by the Rand Corporation in the 1970s. A Delphi study is designed to systematically solicit and collate informed judgments about the future. The Delphi study among REPRO partners was designed to get insight in the areas respondents considered (un)important for the (future) development of REPRO, and in the areas respondents considered (in)feasible for the (future) development of REPRO. The questionnaire consisted of 42 items divided over 6 sections, and was sent to 33 participants of whom 16 responded.

Topics that showed diverging viewpoints (majority versus minority views) are concerned with the relative importance of developing traceability systems of the two trajectories within REPRO, stronger focus on legal compliance, as well as whether REPRO should put more emphasis on developing products with health benefits or animal feed products.

With regard to traceability, respondents argue that it should be considered as an overarching concept that goes beyond the specific trajectories chosen to work on in REPRO. Respondents' views in this respect are related to the fact that they consider the project to exemplify tools and processes, and set an example, rather than to overemphasize certain elements (such as traceability, or developing HCCPs). However, there are some critical remarks made about the trajectories as such, which counter the notion of the project being an example on many terrains. Critical remarks are also made with regard to the current balance between work on food and feed products.

Legislative compliance is considered important, but respondents seem to be in favour for striking a balance between compliance and flexibility, in the sense that legal requirements should not hinder exploratory research in its first stages. With regard to developing products with health benefits, the majority points out that this would result in applying a too narrow focus on the project, to the extent that it's main goals (reduction of waste) could be threatened.

Interestingly, respondents also differ to the extent to which they want to provide information to consumers before possible end products are available in the shops. Also, they point out that consumers should be involved in earlier stages. Items on which respondents lacked 'clear' views (and hence opted for the answer category 'neither/nor') included questions that addressed the feasibility and significance of REPRO's contribution to reprocess food waste and whether the trajectories REPRO is working on (BSG, trimmings) are best suited to reach the project's goals. Similar to what was said previously, for some parts REPRO can be considered a project which aim is to exemplify processes and tools, but for another part the actual significance or feasibility cannot be determined yet. Critical remarks are made with regard to whether the trajectories are best suited to reach the project's goals, and to feasibility in terms of economical feasibility (the costs involved).

3.2 Delphi study among Stakeholders

In the Delphi study among stakeholders we used almost the method as in the Delphi study among Repro members. The differences were that we used ranking questions as well, the "no opinion" option was left out, and the questionnaire was restricted to 16 questions. From the 54 invited candidates only 16 eventually responded.

The main result was that the diverging viewpoints were concerned with the relative importance of communication towards consumers and product-origin-information, the relative importance of increased quality and hazard control and, the appropriateness of existing reprocessing technology. Food safety and quality control was seen as the core issue in relation to consumer trust and acceptance, but communication was considered important by others as a precaution to win trust and handle negative reactions. The idea for consumers to know that food was produced from waste was seen as rather threatening, while reprocessing itself was not seen as risky, just uneconomical for the moment. This in contrast to remarks that the origin of the by product itself plays an important role. Reprocessing inside the production chain is seen as more acceptable than outside, reprocessing vegetable matter more consumer acceptable than animal products.

Legislation and quality control was not perceived to cause greater waste streams, although it alternately was regarded as an obstacle to further utilize by-products. Waste handling legislation was also indicated to form a possible incentive in stimulating reprocessing techniques. Where some reported a lack of technology for reprocessing, others indicated economic incentives were missing to employ existing known reprocessing technology. Others again were pointing out that the existing techniques were lacking an outlook for profitable and acceptable products. Overall in the study, it appeared that although there was sufficient interest in discussing quality aspects of reprocessing of food products, the stakeholders in majority choose techniques for volume reduction and water recycling –not resulting in food products- as the most promising areas for reprocessing.

4 FOCUS GROUP DISCUSSIONS

Focus groups are used to discover what people think and feel with the intention to define the concept of satisfaction, identify the relevant ingredients of satisfaction and discover the conditions or circumstances that influence satisfaction. Characteristics of a focus group: people who possess certain characteristics and provide qualitative data in a focused discussion to help understand the topic of interest. Four focus groups were conducted of respectively 8, 10, 10 and 9 respondents, a total of 37 respondents. After completing the fourth focus group theoretical saturation was reached: the range of ideas was heard, answers overlapped. Moreover, it was not doable to conduct more focus groups in the Netherlands as well as any abroad with the time and resources available. The duration of the focus group discussions was two hours. The audio tapes of the discussions were transcribed and analysed.

In general the participants consider reprocessing as only one step, and not the most important one, in a wider process where waste volumes can be reduced. Taking responsibility over the environment and aiming for sustainable production by buying reprocessed ingredients, has a solid support base from consumers on condition that the industry takes its own responsibility in prevention measures as well. Reprocessing should primarily serve the goal of waste volume reduction and not only serve the economic interests of thc industry. High quality ingredients will keep a second grade image when produced from residue flows. Consumers are not prepared to pay a premium for them. Additional attributes are not sought after, the main goal, a sustainable environment, is the most important. However they are welcomed as a positive side effect and could contribute to acceptability. There are only moderate concerns about food safety. General standards are accepted as sufficient and it is believed that regulations and producers interest are sufficient tools to assure quality and safety. There is only a moderate desire to be informed about the reprocessing of residues into ingredients. It is accepted that we do not know much about most food products. Detailed information about the nature of the used residues and the reprocessing technique are hardly appreciated, but should be available to the ones looking for it. There is a wide recognition that by-products with valuable substances in them should not be regarded as or named "waste". Once turned to good use there is a sense that these are no different resources for production then others. Semantics are seen as important. There is recognition that at the rational level word choice does not make a difference, but that spontaneously, a word as "waste" does trigger different associations than a word as "by-product".

Brewers Spent Grain (BSG), one of the two examples of waste studied in REPRO, ends low on the scale of suitability for reprocessing as compared to other by-products, such as trimmings and rejected vegetable supplies. BSG has associations with bad smell, dirtiness and unhygienic, and is not something consumers easily identify with. For consumers it is hard to imagine possible products. As it is assumed that BSG under new regulations would not be permitted as animal feed anymore, food quality and safety for human consumption are brought forward as areas of concern. Extracting fibre from BSG is thought to deliver a significant volume reduction, which is appreciated by consumers. Assurances are sought around the lead time and storage conditions before reprocessing, to control optimum quality of the by-product. Also it is indicated that reprocessing is required and should be such that the product is freed of all possible contaminants.

Consumers have no real objections against ingredients made from BSG, but there is a general feeling that one wouldn't want to know about their origin. Detailed information is hardly appreciated, but should be available to the ones looking for it. It is also felt that acceptance would be easier if the ingredients are used in existing, already familiar, food products. Reprocessed BSG is perceived as second grade and consumers are thus not prepared to pay a premium for them. Additional attributes, such as the healthiness would be welcomed as a positive side effect and could also contribute to acceptability. Our conclusion is that BSG reprocessing is acceptable under conditions.

Red cabbage trimmings, the second example of waste studied in REPRO, are easily to identify with by-products, and disposal of these trimmings is considered wasteful. Therefore, there are hardly concerns about food safety. General standards are accepted as sufficient on the assumption that regulations and producers' interests are sufficient tools to assure quality and safety. Reprocessing red colorant from red cabbage however, is thought to have only a marginal impact on waste volume reduction. We found that reprocessing should primarily serve the goal of waste volume reduction and not economic interests.

Consumers not only question how much waste is left after reprocessing, but also question whether other options can reduce the waste volume further. Even high quality ingredients are perceived as second grade when produced from residue flows (or "waste") and consumers are thus not prepared to pay a premium for them. Additional attributes, such as the healthiness of the red colorant, are not sought after specifically, but are welcomed as a positive side effect and could contribute to acceptability. For consumers the main goal, a sustainable environment, remains the most important. The ideas of the consumers about information are the same as in the case of BSG. Our conclusion is that red cabbage colorant extraction does not answer consumer expectations.

5 COST BENEFIT ANALYSIS

5.1 Brewers Spent Grain

All over Europe the consumption of extruded snacks is high. The main ingredient of these snacks is rice flour. Research in REPRO[5] has proven that in a chickpea snack it can partly, up to 25%, be replaced by brewer's spent grain.[6] We compared the BSG chickpea snack with a chickpea snack without the dried BSG, but instead containing a higher amount of rice flour.[7] The costs for handling losses of product during production were assumed to be negligible and were thus set to zero. We assumed that the costs for labour, equipment, management, capital, and buildings are the same in both cases. Only the costs for ingredients and the process costs of both kinds of snacks were compared. It was impossible to calculate the benefits of a BSG chickpea snack because the price of the BSG chickpea snack was unknown, and there is no product yet on sale in the market. Also prices may vary when we take into account whether there is a health effect from the dietary fibre in the BSG chickpea snack.

The reference product contains 21% BSG on a dry matter basis. The total cost of 92.6 kg chickpea snacks is €7.69, comprising the cost of the rice flour (€1.16) and the costs of the other ingredients including processing (€6.53). The total cost of 92.6 kg BSG chickpea snacks is €18.30, comprising costs for other ingredients and process costs (€6.53), the cost of gum arabic (€11) and the cost of additional energy (€0.77). A major difference between rice flour and BSG flour is that wet BSG needs to be chilled and dried, in addition of milling. The extra energy cost per kg dry BSG is €0.77, of this amount €0.65 is for drying in the oven and €0.12 is for chilling. Transport distances, based on simulations of 50 and 1000 km, are not critical with regard to costs. For the BSG chickpeas snack, the BSG is almost free, whereas, for the chickpea snack, in addition to the other ingredients, 21% more rice flour has to be bought.

Compared with the case of the chickpea snack, the costs for gum arabic and for the drying step more than doubled the costs in the case of the BSG chickpea snack. It is worth noting that, although gum arabic is only a minor ingredient, it nevertheless dominates the production costs. To reduce the overall production costs, existing cheap alternatives to gum arabic may be considered. The drying step is also a relatively expensive extra step. Costs can be decreased by pressing the BSG before processing and possibly by drying the product at the brewery using other energy sources e.g. excess steam. When gum arabic is replaced by a cheap alternative, the total assessment of the BSG chickpea snacks will be positive. Transport has only a small effect on cost.

5.2 Red cabbage trimmings

When cabbage is processed, a certain percentage is lost as trimming. By using enzymes, red cabbage trimmings are partly degraded and valuable components can be extracted .[8] Research in REPRO achieved the following products: pectin, 'modified hairy regions' (MHR), (this is a health promoting substance[9], dietary fibre, and a phytochemical rich cabbage juice. The costs versus benefits were evaluated for two scenarios: the production of pectins and MHR, and for the production of only MHR.[7] Costs per day (per 10 tonnes) arise from resources (enzymes, buffers, and water), electricity, natural gas, and transport. Experts in the REPRO project estimated that the benefits per day (per 10 tonnes) could be: natural colorant (€50 per kg), MHR (€7 per kg), pectin (€7 per kg), and fiber (€1 per kg). In both cases (with or without pectins) the costs per day for labour, capital, management, buildings, and equipment are left out of the calculations as it is unknown if it is possible to incorporate the cases into an existing production facility. For this reason our recommendations are only based on the process costs.

In the best case of the pectin with MHR scenario, the benefits of €1942 minus the process costs of €3856 show a loss of €1914. Without even taking into account the other costs we can conclude that the total cost/benefit assessment for this case is negative. In the best case MHR alternative scenario, the benefits of €1872 minus the process costs of €1381 show a gain of €491, but there will also be other costs. For this reason the total cost/ benefit assessment is only relatively positive. After the inclusion of the other daily costs, the total cost/benefit assessment of this scenario will also be negative. Energy efficient drying, including pre-concentration, will substantially contribute to cut costs and lower the environmental impact of the process. Alternative energy resources to electricity should be used as far as possible. For the optimized process (reference scenario), the costs of enzymes and energy are dominant.

6 CONCLUSION

An important result of our studies in the context of REPRO is that stakeholders and consumers acknowledged the need of reduction of waste. However, these groups also doubted whether the focus on novel and functional food ingredients is the best way to reduce waste volume. Reducing food processing waste is placed in a broader spectrum of issues, such as the amounts of food waste produced at the household level and supermarkets. It is not clear how cost effective it is to develop new products in relation to other solutions that reduce the amount of co-products left unused, e.g. composting or biodegradable packaging. Retailers differentiated between the generally acknowledged need to reduce the amount of food waste, but pointed to several barriers that may prevent successful implementation. These vary from practical considerations to broader concepts such as psychologically and/or culturally determined barriers to reprocess waste or co-products. They believed the way in which information is presented influences respondents' opinions. They reacted very strongly to the wording of certain REPRO texts. REPRO partners pointed out that the development of products with health benefits could result in applying a too narrow focus on the project, to the extent that its main goals (reduction of waste) could be threatened. This was acknowledged by consumers. With consumers, the need to reduce the amount of food processing waste is generally acknowledged and presented in terms of a strong normative obligation. They were questioning the validity of reprocessing if volume reduction was not its main goal.

Industrial stakeholders saw food safety and quality control as the core issues in relation to consumer trust and acceptance, but communication was considered important as a precaution to win trust and handle negative reactions. Both stakeholders and REPRO partners differed amongst each other in the extent to which they wanted to provide information to consumers before possible end products are available in the shops. For consumers trust was mainly put in the marketing interests of producers and retailers. There was little interest in additional information, but an express demand that information should be available and transparent for those who were interested. Consumers distinguished between processing technologies used to develop novel food products and possible end products. They were more critically aware of the need for a transparent production process than the extent to which end products such as juices or snacks will be successfully sold in shops. Communication was seen as both required and threatening at the same time by REPRO partners and stakeholders. We found that trust and responsibility, the producers' reassurance and control of market risk factors were important consumer issues. Among consumers there was a clear interest in the quality of the by-product prior to reprocessing, and they saw a direct relation between the quality of the by-product and the quality of the reprocessed ingredient. The perception of quality was strongly linked to everyday practices.

Brewers Spent Grain ends low on the scale of suitability for reprocessing as compared to other by-products, such as trimmings and rejected vegetable supplies. BSG has associations with bad smell, dirtiness and unhygienic, and is not something consumers easily identify with. Consumers had no real objections against ingredients made from BSG, but there was a general feeling that one wouldn't want to know about their origin. It was felt that acceptance would be easier if the ingredients would be used in existing already familiar food products. Reprocessed BSG was perceived as second grade and consumers were thus not prepared to pay a premium for them. Additional attributes, such as the healthiness would be welcomed as a positive side effect and could also contribute to acceptability.

Red cabbage colorant extraction did not fully answer consumer expectations. The trimmings were easily accepted as a resource for further processing. Therefore, there were hardly any concerns about food safety. However, reprocessing red colorant from red cabbage was thought to have only a marginal impact on waste volume reduction. Consumers questioned how much waste will be left after reprocessing. Consumers further questioned whether other options could reduce the waste volume further. Also, even high quality ingredients were perceived as second grade when produced from residue flows (or "waste") and consumers were thus not prepared to pay a premium for them. Like in the case of reprocessed BSG, additional attributes, such as the healthiness of the red colorant, are not sought after specifically, but were welcomed as a positive side effect and could contribute to acceptability.

References

1 K. Waldron *et al.*, *Reducing Food Processing Waste,* REPRO, final report, EU project no. Food-CT-2005-006922

2 L.J. Frewer and B. Gremmen in *Handbook of waste management and co-product recovery in food processing, Volume 1*, ed. K.W. Waldron, Woodhead Publishing, Cambridge, UK, 2007, p 21.

3 J. Baudrillard, 1998, *The Consumer Society*, Sage Publications, London, 1998.

4 M. Douglas, *Purity and Danger: An analysis of the concepts of pollution and taboo*, Allen and Unwin, London, 1966.

5 P. Ainsworth, Ş. İbanoğlu, A. Plunkett, E. İbanoğlu and V. Stojceska, *J. Food Eng.*, 2007, **81**, 702.
6 Ş. İbanoğlu, P. Ainsworth, E.A. Özer and A. Plunkett, *J. Food Eng.*, 2006, **75**, 469.
7 B. Gremmen in *Handbook of waste management and co-product recovery in food processing, Volume 2,* ed. K.W. Waldron, Woodhead Publishing, Cambridge, UK, 2009, p 23.
8 J. Schroot, A. Zykwinska, M. Panouille, S. Hinze, G.I.A. Borge, E. Bonnin and H.A. Schols, Recovery of valuable compounds from red cabbage trimmings, 2009, *TDC-OLIVE. By-product reusing from olive and olive oil production*, biomatnet, 1859.
9 H.A. Schols, M.A. Posthumus and A.J.G. Voragen, *Carbohydr. Res.*, 1990, **206**, 117.

Value Added Products - Plants

FOOD APPLICATIONS OF NOVEL INGREDIENTS FROM AGRO-BASED SUSTAINABLE SOURCES

J.L. Bialek, D. Jarvis and P. Lopez-Sanchez

Unilever R&D, Olivier van Noortlaan 120, 3133AT Vlaardingen, Netherlands

1 INTRODUCTION

Millions of tonnes of vegetables and fruits such as onions, tomatoes, strawberries are grown in European countries and, a large proportion of them are used in processed food. This yields vast quantities of waste rich in insoluble and soluble solids (pieces of tissues, cell wall polymers, and endogenous enzymes). As an example, within Europe alone, over 500.000 tonnes of onions are left as waste material annually, leading to a large potential source of low cost material. Furthermore today's more and more health conscious consumers have a strong demand for more "natural," minimally processed, additive free foods. Therefore, on environmental and economic grounds, the food processing industry feels the pressure to explore and exploit the potential of plant material components as a structuring material for various food products. In this context, many studies were carried out on properties of vegetable pastes, especially on their rheological properties as gelling agents or thickeners. For instance Xu[1] studied the flow behaviour of tomato pastes, whereas Thibault[2] explored the gelling properties of extruded citrus fibres. In this work, it has been observed that mixing Thermally Treated (TT) and Non-Thermally Treated (NTT) fruits or vegetables gives rise to gelation. The addition of such a mixture to liquid or pourable products shows a strong thickening effect.[3] In the study reported here, the tissue of interest is onion (*Allium cepa* L.). Onions have attracted attention because of their application in various food products like sauces and dressings. Two different processes define Thermally Treated (TT) onion purée and Non-Thermally Treated (NTT) onion purée. Heating to 90 °C and then breaking the tissue of the vegetable produces Thermally Treated onion purée, whereas in the case of Non-Thermally Treated, the flesh is broken without applying heat. Heat treatment during processing is known to lead to an inactivation of endogenous cell wall degrading enzymes, such as polygalacturonase or pectin-methyl esterase (PME).[4] It also causes changes in the solubility, size and charge density of the cell wall pectins.[5,6] Pectins are complex mixtures of polysaccharides, which constitute approximately one-third of the dry weight of higher primary cell walls[7]. These polymers are composed of linearly connected α (1-4) D-galacturonic acid residues randomly interrupted by 1-2 linked rhamnose residues.[8] Based on the degree of esterification (DE), which is the proportion of carboxyl groups present in the esterified form, pectins with a DE lower than 50% are called low-methoxyl (LM) pectins, while those with DE equal to or greater than 50% are called high-methoxyl (HM) pectins. The mechanism for gel formation is different in both types of pectins. The gel formation of HM pectins requires conditions of low pH (~3) and low water activity (typically, high sugar content ~65%), and is governed by both hydrogen bonding and hydrophobic interactions.[9] In contrast, LM pectins form gels in the presence of calcium ions over a wide range of pH even without added sugar. The "egg-box" model, that has been widely accepted for calcium pectate gels[10], describes the junction zones in a LM pectin gel as the consequence of the interaction between the calcium ions and the carboxyl groups of the pectin chains.[11] Pectin methylesterase (PME) catalyses the removal of the methoxyl groups from the pectin molecule in a sequential fashion, creating blocks of free carboxyl groups and releasing protons and methanol.[12] The reaction of this plant enzyme is found to be quite specific, requiring at least one galacturonic acid residue free of methoxyl groups in order to attack an esterified unit.[13]

Onion purée is composed of two distinct phases; (1) a continuous liquid phase (the serum) and (2) an insoluble solids phase made up of plant cell wall fragments. From the work of Gothard and Whiteman, it has been shown that addition of PME extract to thermally treated tomato paste induced thickening and gelation. The proposed mechanistic model for thickening and gelation of PME treated tomato pastes highlighted the importance of the soluble pectic polymers present in the serum phase for PME mediated pectin structuring. From this above work, it seems justified to assign the gelation observed for the onion purées mainly to the PME modified pectins of the cell wall. In this model, pectin methyl esterase from Non-Thermally Treated serum de-esterifies HM soluble pectins from Thermally Treated purées, generating free carboxyl groups and thus soluble LM pectin. The exposed carboxyl groups can cross-link with other regions of de-esterified galaturonic acid residues by means of calcium ions. The creation of "egg-box" junction zones leads to polymer network development, which may ultimately lead to the formation of a gel. However, this does not exclude a contribution from the insoluble solid phase on the resultant gel strength. The purpose of this study was to test the onion gelation model by investigating the relationship between rheological properties and the microstructure of TT/NTT onion mixtures. The role of serum, insoluble solids, the NTT / TT phase volume and particle size were tested for their effect on gel strength.

2 MATERIAL AND METHODS

2.1 Preparation of Thermally Treated and Non-Thermally Treated onion puree

Onion (*Allium cepa* L., Hysam variety) was obtained from a local onion producer after commercial harvesting (G's Marketing Ltd, Cambs, UK) and were stored at 4°C. The brown dry outer skin and the top and bottom ends of the onions were removed and discarded. The fleshy scales of the onions were then cut into pieces 1cm × 1cm × natural thickness in size. Two different methods were used to obtain Non thermally Treated (NTT) and Thermally Treated (TT) purées.

Onion purées were prepared using a Kenwood FP570 food processor with either the main processing unit equipped with a twin metal blade (>300 g batch) or the liquidiser accessory (≤ 300 g batch).

2.1.1 Non Thermally Treated onion purées (NTT). 300 g of onions, were processed for 2 minute 30 seconds on minimal speed (level 1) in the liquidiser.

2.1.2 Thermally Treated onion purées (TT). 300 g batches of freshly chopped onions were placed into a microwave (1925-Watt, Merrychef Turboaire II) and heated for 2 minutes on full power. On exit from the microwave the temperature of the onions was checked to be at least 90 °C using a temperature probe. The temperature of 90 °C was chosen according to previous work to ensure de-activation of enzymes.[12] The onions were re-weighed and de-ionised water was added to replace the water evaporated during cooking. The cooked onions were then broken in the food processor using the twin metal blade, 20 minutes per 1000g on maximal speed.

NTT and TT purées were forced separately through a 2 mm or 1 mm brass sieve to reduce the flesh to a standardised particle size. Samples were stored in airtight containers at 4 °C.

2.2 Separation of insoluble solids from serum and preparation of serum and pellets (insoluble solids) mixtures

To separate the insoluble solids and serum a weighed amount of NTT and TT onion purées, (sieved < 2 mm) was centrifuged at 20 °C for 20 minutes at 10 000 rpm and 14,000 rpm respectively using a Sorvall RC 5C centrifuge (SS-34 rotor). The supernatants were passed through a filter Miracloth to take only the liquid serum phase off. Serum and pellets were weighed and were retained for further experiments.

2.2.1 Preparation of the serum mixture. The NTT and TT serums obtained from centrifugation of the onion purées were mixed together in the desired (w/w) ratio. NTT and TT serums and their mixtures were incubated at 30°C for 3 days.

2.2.2 Preparation of the pellet (insoluble solids) mixture. The pellets were separated from the serum after the centrifugation step. The serum was decanted and replaced by an equal volume of deionised water. The "new purées" obtained from the thus re-suspension NTT and TT pellets were then used to create mixtures of NTT and TT in the desired (w/w) ratios.

2.3 Light Microscopy

Bright field micrographs of onion purées were obtained using a Leica Polyvar light microscope with ×4 objective and ×0.8 or ×1.25 zoom for either cells pictures or background pictures respectively. Images were recorded in real colour and onion purées were diluted to 5% (v/v) with de-ionised water before imaging.

2.4 Rheological measurements

Dynamic rheological tests were conducted with a Paar Physica UDS 200 rheometer in the oscillation mode. SAOS (Small-Amplitude Oscillatory Shear) experiments were performed by applying a small sinusoidal shear stress. From the different geometries tested a combination of vane (12.5 mm) and serrated cup gave the best reproducibility of results, especially for NTT onion purée, and was thus used in all experiments.

Gelation and especially the gel strength was followed by recording the storage modulus (G') as a function of time. The time sweeps were performed during a period of two hours unless otherwise indicated, at a 1Hz constant frequency and 1 Pa constant stress. This value was selected by a stress sweep test to guarantee a linear viscoelastic response.

The microstructure of samples containing pellet re-suspended in deionised water was measured by recording the storage modulus G', and the loss modulus G" as a function of frequency. The frequency sweep applied was a ramp log frequency 0.2Hz - 10Hz with 1 Pa constant stress.

After loading 40g of onion purée into the serrated cup the exposed sample surface was covered with a thin layer of mineral oil to avoid water evaporation during rheological tests. The experiments were performed at 30 °C unless otherwise indicated.

3 RESULTS AND DISCUSSION

3.1 Phase volume effect

In the first place it was very important to establish how changing the phase volume ratio of NTT/TT may affect the structuring properties of the onion mixture. The structure formation of the onion mixture was studied using oscillatory measurements in the time sweep mode and storage modulus – G'was used in comparisons.

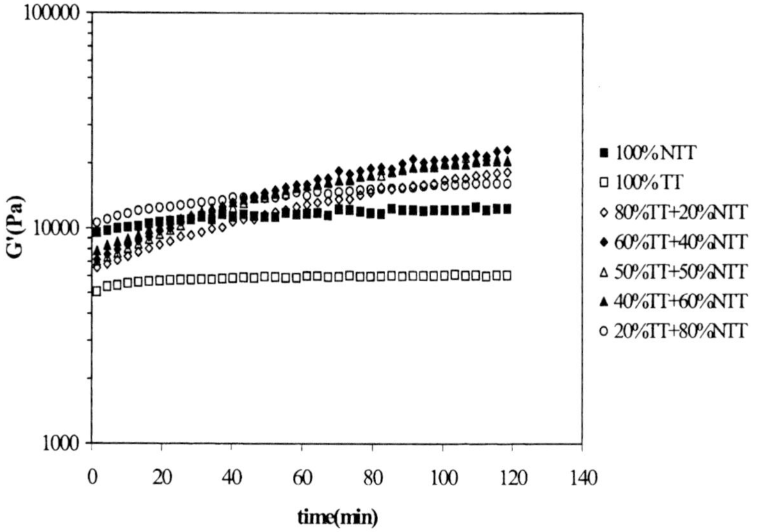

Figure 1 *Phase volume effect: Conditions: % represents (w/w) ratio, purees sieved (<2 mm). Conditions: Time sweep (1 Pa shear stress, 1Hz frequency), duration 2 hours, temperatue 30 °C.*

Figure 1 shows development of G' (storage modulus) as a function of time for various ratios of TT and NTT onion purées. For 100% NTT or 100% TT onion purées no significant change in G' over the incubation period was observed. This suggests that neither NTT nor TT onion purée was capable of gelling. The G' value of NTT was approximately 4000 Pa greater than that of TT. On the contrary, mixing NTT/ TT onion purées at different phase volume resulted in significant increase of G'. After 2 hours of measurement, the highest G' was observed for 60 %(w/w) TT + 40%(w/w) NTT onion purée mixture. After 2 h of oscillatory measurements the G' value seems to reach a plateau. A 2hr-time scale was used for the further study of the particle size effect on the gel formation.

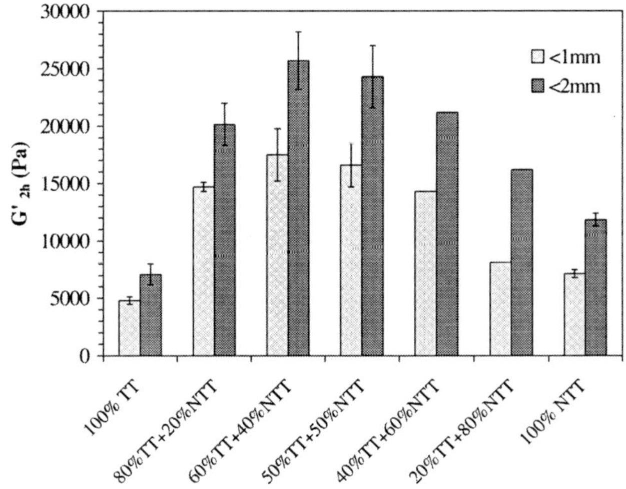

Figure 2 *Phase volume and particle size effect (insoluble solids) on the resultant gel strength (G'_{24}). Conditions: Time sweep (1 Pa shear stress, 1Hz frequency), temperature 30°C.*

Figure 2 shows both the effect of changing the NTT/TT ratio and the effect of reducing the particle size. The results are presented as a function of elastic modulus after 2hrs of incubation at 30°C. Error bars indicate a relatively large variability of the measurements. This could be due to both environmental (change in temperature, humidity, etc.) and equipment related problems, but it is clear that the nature of the raw material is the main reason. Studies of living material, such as vegetables, are usually subject to this variability problem. As a consequence, our analysis focuses more on trends than absolute results.

The same general trends were observed for both particle sizes considered in this study. NTT onion purée had a greater storage modulus than its TT counterpart and the largest value of G'_{2h} was obtained for 60% (w/w) TT + 40% (w/w) NTT onion mixtures. As shown in Figure 2, a small amount of NTT onion added to TT onion purée was sufficient to obtain a significant increase in G'_{2h}, whereas in the opposite scenario this trend was less pronounced. From these observations, it seems that NTT/ TT onion purées act on the resultant gel strength differently. The reduction of particle size results in a decrease of the G'_{2h} values for all ratios studied. However this decrease was less significant for 100% TT onion purée (about 2000 Pa) than for 100% NTT onion purée (about 5000 Pa).

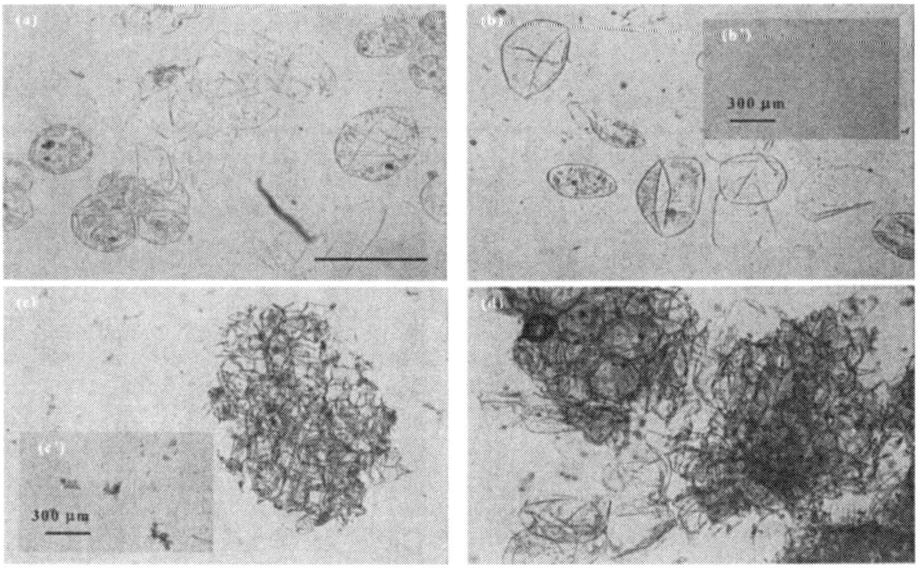

Figure 3 *Brightfield images of onion purees: (a) Cells of TT onion puree sieved <1 mm; (b) Cells and (b') background (serum) of TT onion puree sieved <2 mm; (c) Cells and (c') background (serum) of NTT onion puree sieved <1 mm; (d) Cells of NTT onion puree sieved <2 mm. Bar scale = 500 µm.*

The images focused on the microstructure differences between TT and NTT onion purees. Figures 3a, b & b' presents the microstructure of TT onion puree. TT onion puree contains many intact cells and, smaller cell clusters. Two distinct distributions of the particle have been observed. The first group (Figure 3a, 3b) is in the size range between 100 – 400 µm diameter. The second group of particles are much smaller than the first one, consisting of fine homogeneous debris of cell walls and nuclei (Figure 3b'). These are in a size range of less than 15 µm. As for NTT onion purée (Figures 3c, 3c'&d), the background and cells seem more heterogeneous than ones of TT. Despite this heterogeneity, two particle size populations could be observed. First population is composed of almost clumps of cells with some intact cells in inner, some pieces of tissues, and a few "cell ghosts." Pieces of tissues and clumps are in a size range more than 1 mm (Figures 3c & d) whatever purée, sieved less than 2 mm or 1 mm, however, they are more numerous and bigger in the former. Debris, collected and clumped residues from broken cells form the second heterogeneous population (Figure 3c'). All of these particulates are suspended in a solute phase, the serum.

3.2 Contribution of Serum and Insoluble solids to rheology of onion mixture

In order to determine the relative contributions of insoluble solids and serum fractions on purée rheology and the gelation of different mixtures, purées were centrifuged to separate serum from pellet. Then various serum replacement and serum/ pellet mixing experiments were carried out.

3.2.1 *Role of serum polymers.* Serum gelation was studied by visual observation and the time sweep measurement.

Table 1 *Serum gelation after incubation at 30 °C for 3 days based on visual observation.*

% correspond to w/w ratio	Visual observations
100 % TT serum	**no gelation**, light yellow opaque **viscous liquid**
80%* TT serum + 20% NTT serum	**gelation, weak** continuous gel
60% TT serum + 40% NTT serum	**gelation**, continuous gel **stronger** than 80% TT/ 20% NTT gel, free water at the surface
50% TT serum + 50% NTT serum	**gelation**, gel with **large aggregates**
40% TT serum + 60% NTT serum	**gelation**, gel with **medium aggregates**
20% TT serum + 80% NTT serum	**gelation**, gel with **small aggregates**
100% NTT serum	**no gelation**, pink clear very **low viscosity liquid**

The serum contains small solutes (e.g. sugars, salts) and either HM pectic polymers or PME for TT and NTT serum respectively. When TT serum and NTT serum were mixed and incubated at 30 °C for 3 days a gel resulted. The strongest gel was obtained from 60% TT + 40% NTT ratio (Table 1). This visual observation suggests that at this phase volume ratio the balance between soluble HM pectin present in the TT serum[14], and PME in the NTT serum is the most desirable to give rise to a continuous weak gel. The reported suggestion fits well with PME mediated structuring previously proposed by Gothard and Whiteman which suggested that the degree of structuring was a function of pectin and PME concentrations.

In order to quantify the results, the time sweep experiment on 60TT/40NTT serums mixture was conducted – Figure 4. The gelation process was slow and started after 170 minutes at 30 ^{0}C. At the end of the incubation time the elastic modulus reached value of 170 Pa. This experiment confirmed gelling capability of the sera mixture.

Viscosity measurement of the NTT and TT serums, using a Contraves low shear 30 rheometer, showed that NTT serum possesses much lower serum viscosity than its TT counterpart, 1.34mPas and 23.3mPas respectively. This is thought to be due to lower concentrations of soluble pectic polymer, but there may also be molecular weight and compositional arguments for the observed differences.

3.2.2 *Role of insoluble solids (pellets).* After structuring capabilities of the serum has been established it was important to determine the rheology of the insoluble solids (pellets) without contribution from the serum phase. The range of NTT/TT pellets ratios were dispersed in deionised water and measured on the Physica rheometer.

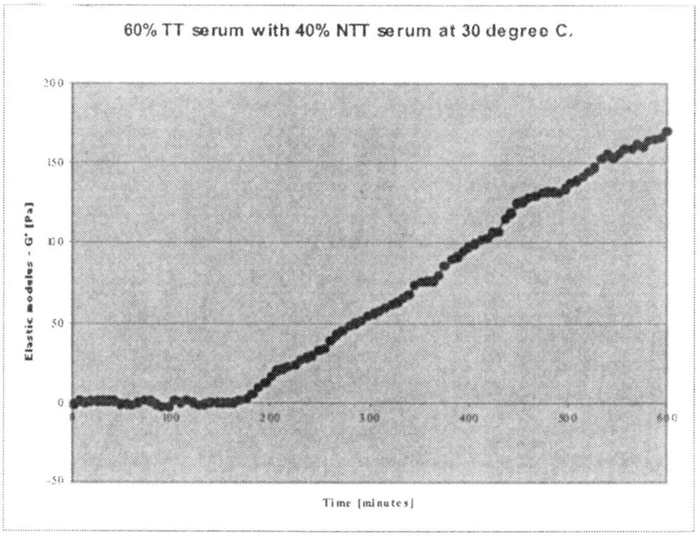

Figure 4 *Time sweep, duration 10 hrs, max G' 170 Pa.*

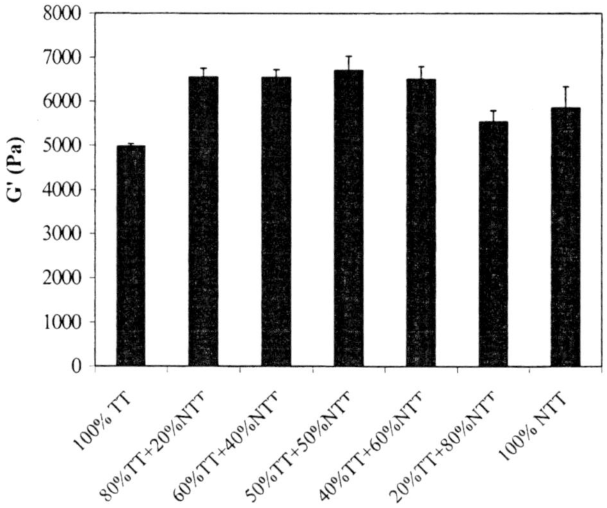

Figure 5 *NTT/TT pellet phase volume effect on the storage modulus. Conditions: Pellets separated from onion puree sieved <2 mm and re-suspended in deionised water. Frequency sweeps (ramp log from 0.2Hz to 10Hz, 1 Pa stress), G'at 1.1 Hz, temperature 30 ^{0}C.*

When TT particles and NTT particles separated from the serum phase were mixed and tested in a rheological frequency sweep, no significant differences appear from G' values (around 7000 Pa) of the different NTT/TT ratio (Figure 5). Several results (Figures 1,2) have revealed that the storage modulus of NTT onion purée is greater than its TT counterpart and that the gel strength of the mixture is closer to NTT purée than TT. This suggests that the solid properties of onion purée mixture are not related to the NTT/TT pellet phase volume but that insoluble solids dominate the rheological behaviour. This assumption is supported by images from light microscopy (Figure 3). Differences of microstructure between TT and NTT onion purées could be explained as follows. For TT onion purée, thermal treatment has induced a softening of onion tissues and therefore the cells separate much more easily under shear, leaving many more intact cells.[15] Heat-induced softening of onion tissue results in the solubilisation of HM pectin, which is accompanied by the loss of pectic polymers from the cell wall. This is probably due to dissolution, depolymerisation, and perhaps to destruction of cell wall pectins through β-eliminative cleavage.[16] In the case of NTT onion purée, PME is loosely bound to the cell wall and physical disruption of the tissue during the breaking process will give it good access to its natural substrate, methyl esterified pectin. Once de-esterified the pectic acid is likely to crosslink via divalent cations such as calcium. If calcium cross linking occurs before the pectin is released from the cell wall environment, then pectin would be bound within the particle or with pectin from other close cell wall material (clump formation). In addition, the microstructure observed for NTT onion particles could be related to the break process itself. NTT onions are broken without applying any heat treatment, this results in the preservation of a large amount of original tissue fragments or aggregates of cells. From the microstructure of NTT and TT onion particles it is obvious that they can not have the same participation in purée rheology. Moreover, it was established that particle size reduction involved a decrease of the final gel strength. This is believed to be due to the dependence between serum: pellet, ratio and NTT particle size. Thus, the former increases when the latter decreases, this then results in a decrease of the amount of insoluble solids in NTT onion. Consequently, particle size effects might be due to the reduction of the amount and size of cell aggregates from the insoluble NTT phase. These arguments support the fact that NTT insoluble solid is the main factor in gel strength. Ionic composition and the mechanical shearing conditions are believed to control the solubilisation of enzymes ionically bound to cell wall fragments. Since experiments always were performed in the same shearing conditions, so, particle size effects are unlikely to be linked to modification of PME availability.

4 CONCLUSION

The above work has demonstrated that onion gelation depends on phase volume of Non Thermally / Thermally Treated onion purée and that the gel strength is affected by particle size of the sieved purées. Gelation of the corresponding mixture is governed by the components present in the serum phase. The action of PME contained in NTT treated onion serum de-esterifies the soluble pectin polymers mainly present in the TT onion serum, permitting formation of a calcium cross-linked network. In the onion model, pectin matrix does not include insoluble solids as a component of it, but rather as inclusions within it, thus increasing its overall strength. In this resultant gel, the NTT insoluble phase more than TT might dominate the final solid properties and could be responsible for enhancing moduli and gel strength.

References

1 S-Y. Xu, C.F. Shoemaker and B.S. Luh, *J. Food Sci.*, 1986, **51**, 399.
2 J.-F. Thibault, M.-C. Ralet, M.A.V. Axelos and G. Della Valle in *Pectins and Pectinases*, ed. J. Visser and A.G.J. Voragen, Elsevier Science Edition, 1996, p425.
3 J. Bialek, I. Appelquevist and C. Watson, 2002. Thickening Agent and process for thickening. Granted patent, International publication number WO 2004/023893 A1
4 B.S. Luh and H.N. Daoud, *J. Food Sci.*, 1971, **36**, 1039.
5 C.Y. Lee, M.C. Bourne and J.P. Van Buren, *J. Food Sci.*, 1979, **44**, 615.
6 N. Ben-Shalom, D. Plat, A. Levi and R. Pinto, *Food Chem.*, 1992, **45**, 243.
7 R.H. Walter in *The chemistry and technology of pectin*, ed. R.H. Walter, Academic Press, San Diego, 1991, p1.
8 J.M. Lau, M. McNeil, A.G. Darvill and P. Albersheim, *Carbohydr. Res.*, 1985, **137**, 111-118
9 D.G. Oakenfull in *The chemistry and technology of pectin,* ed. R.H. Walter, Academic Press, New York, 1991, p87.
10 G.T. Grant, E.R. Morris, D.A. Rees, P.J.C. Smith and D. Thom, *FEBS Letters*, 1973, **32**, 195.
11 M.A.V. Axelos and J.-F. Thibault in *The chemistry and technology of pectin*, ed. R.H. Walter, Academic Press, New York, 1991, p109.
12 T. Sajjaanantakul and L.A. Pitifer in *The chemistry and technology of pectin*, ed. R.H. Walter, Academic Press, New York, 1991, p135.
13 L. Catoire, M. Pierron, C. Morvan, C. Herve du Penhoat and R. Goldberg, *J. Biol. Chem.*, 1998, **273**, 50, 33150.
14 A. Ng and K.W. Waldron, *J. Sci. Food Agric.*, 1997, **73**, 503.
15 S. Lecain, A. Ng, M.L. Parker, A.C. Smith and K.W. Waldron, *Carbohydr. Polym.*, 1999, **38**, 59.
16 L.C. Greeve, R.N. McArdle, J.R. Gohlke and J.M. Labavitch, *J. Agric. Food Chem.*, 1994, **42**, 2900.

IMPROVING THE TEXTURAL CHARACTERISTICS OF BREWER'S SPENT GRAIN BREADS BY COMBINATION OF SOUR DOUGH AND DIFFERENT ENZYMES

V. Stojceska and P. Ainsworth

The Manchester Metropolitan University, Department of Food and Tourism Management, Hollings Faculty, Old Hall Lane, Manchester, M14 6HR, UK. Corresponding email: V.Stojceska@mmu.ac.uk

1 INTRODUCTION

Cereal co-products such as brewer spent grain (BSG) have strong potential to be incorporated into bread making technology. BSG significantly improved the dietary fibre level in conventional baked breads but the major difficulty encountered was to achieve a good structure and high loaf volume.[1] Recent findings indicated that the textural characteristics of BSG breads could be improved with the incorporation of different enzymes such as Lipopan, Pentopan Mono and Celluclast. Also, a number of researchers improved the texture, volume and shelf life, nutritional and organoleptic properties of rye and wheat breads by combination of sour dough fermentation and different enzymes.[2,3]

The objective of this research was to study the effect of sour dough and a number of different enzymes on wholemeal BSG breads.

2 METHOD AND RESULTS

The ingredients used for the bread making were: wheat flour (Smiths Flours Mills, Worksop, UK), yeast (Fermipan, Gist-Brocades, Holland), shortening (Cardowan Creameries Ltd., Glasgow), sugar and salt (purchased from a local supermarket in Manchester).

A commercial sample of brewer's spent grain with composition of 75% moisture, 1% ash, 4.8% protein, 16% fibre, 2.1% fat and 1.1 % carbohydrates content was supplied by the Joseph Holts Brewery (Manchester, UK).

Frozen BÖCKER TK Starter (Ernst BÖCKER GmbH and Co. KG, Germany) was used for preparation of sour dough.

Four commercial enzymes were used: α- amylase Grindamyl Max-Life 85 (100-300 ppm), containing 100 Fungal Amylase Units/gram (Danisco Ingredients, Denmark) (ME); Lipopan Xtra BG enzyme (20-30 ppm), containing 7.2 Kilolipase Units/gram (LE); Pentopan Mono BG (20-120 ppm), containing 2500 Fungal Xylanase Units(Wheat)/grams (PE) and Celluclast L (10-60 ppm), containing 700 Endo-Glucanase Units/gram (Novozymes, Denmark) (CL).

2.1 Preparation of BSG

The source of BSG was based on brewing of barley hops and is the remains after wort mashing and straining, before fermentation of the beer. The BSG was refrigerated immediately after collection and then dried the following day at 150°C for 4 hours in a Teknotronic reel oven (Teknigas Ltd, Sussex, England) to a moisture content of 5-7%. The dried sample was milled using 0.5 mesh screens, packed and then sealed in polyethylene bags and kept at room temperature until use. The composition of dried BSG was 20.30% protein, 53.39% fibre, 8.32% fat and 10.76% carbohydrates.

2.2 Preparation of sour dough breads

The dough was prepared with mixing sour dough, soaker dough, wheat flour, water, salt and yeast. Sour dough was prepared with mixing BÖCKER TK Starter, wheat flour and water. Soaker dough was made by mixing BSG and water. Mixing of the ingredients was carried out in a Hobart NCM mixer (Process Plant and Machinery Ltd., UK) at 1st speed. The resulting dough was divided into three 300g pieces, moulded by hand and placed into pre-greased 454g tins. The dough was proofed 55 min at 30°C and 85% humidity and baked in a Tektronic reel oven (Teknigas, Ltd., Sussex, England) for 20 min at 230°C. The following bread types were studied:

- Three control sourdough breads (SBB) with the substitution levels of BSG at 0, 10, 20 and 30%.
- Three sourdough breads using Lipopan Extra enzyme (SLE) with the substitution levels of BSG at 0, 10, 20 and 30%.
- Three sourdough breads using Pentopan enzyme (SPE) with the substitution levels of BSG at 0, 10, 20 and 30%.
- Three sourdough breads using a mixture of Pentopan + Celluclast enzymes (SPCE) with the substitution levels of BSG at 0, 10, 20 and 30%.
- Three sourdough breads using a mixture of Lipopan Extra and Maxlife 85 enzymes (SLME) with the substitution levels of BSG at 0, 10, 20 and 30%.
- Three sourdough breads using a mixture of Pentopan and Maxlife 85 (SPME) enzymes with the substitution levels of BSG at 0, 10, 20 and 30%.

2.3 Bread evaluation

Quality analysis of fresh bread samples was carried out by measuring weight, volume (determined by seed displacement in a loaf volume meter) and specific loaf volume.

Crumb firmness was measured using a texture analyzer (TA-XT2i, Stable Micro Systems, Surrey, UK) at 1, 5 and 8 days to assess potential shelf-life of the breads. Bread crumb firmness during storage was determined as maximum compression force[4]. The loaves were sliced by hand into equal slice thickness of 25 mm thick. The crust slices of the loaves were discarded. Nine bread slices originating from three loaves were measured and results were expressed as mean values.

2.4 Statistical analysis

Statistical analyses were conducted using a multiple comparison test using SPSS 12.0.1 (SPSS Inc., Chicago, Illinois, US). All the experiments were replicated, so that the data in the paper are all mean values of three tests. The coefficient of variation (CV) of all the tests was lower than 10%.

2.5 The effect of sourdough fermentation and enzymes on BSG breads

The specific loaf volume of sourdough breads varied between 2.0-2.5 ml/g for the control breads and 1.8-2.5 ml/g for enzyme added breads (Figure 1). The significant decrease ($P<0.05$) in loaf volume was found at the 30% level of BSG. The addition of enzymes did not improve the loaf volume of sourdough breads.

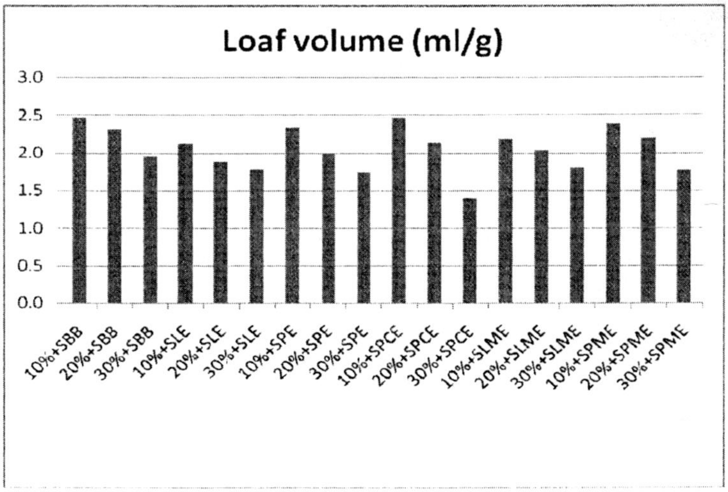

Figure 1 *Specific loaf volume of BSG sour dough breads with a range of different enzymes.*

Figure 2 presents the hardness of the breads observed at 1, 5 and 8 days of storage. The results indicated that increased level of BSG (0-30%) significantly ($P<0.05$) increased the hardness in SPE, SPCE and SPME breads. It was apparent that addition of enzymes improved the texture and shelf life of BSG breads resulting with lower crumb hardness and delay in staling comparing with their equivalent SBB samples. The positive influence of enzymes was more evident in SPCE, SLME and SPME probably as a result of combination of PE and LE with ME and PE with CL, which produced breads with significant ($P<0.05$) decrease in hardness and extended shelf life comparing to the equivalent samples such as SLE and SPE. Use of sour dough and subsequent increased acidity in dough probably improves protein solubility and proteolytic activity, which modifies the gluten network, affecting the physical properties of gluten and the texture of breads.[2]

The results were compared with the recently published work,[1] where the same enzymes were used in improving conventional breads. It was found that sour dough extends shelf life in high fibre BSG breads comparing to conventional BSG breads. The same attitude was more pronounced with a combination of sour dough and enzymes. However, the best results in terms of extending shelf life were obtained with the combination of LE and PE with ME, and PE with CL, which is probably a result of reduced starch retrogradation rate, slowed increase in rigidity of polymer structure and due to degradation of cell wall components leading to a redistribution of water between starch-protein matrix.[3]

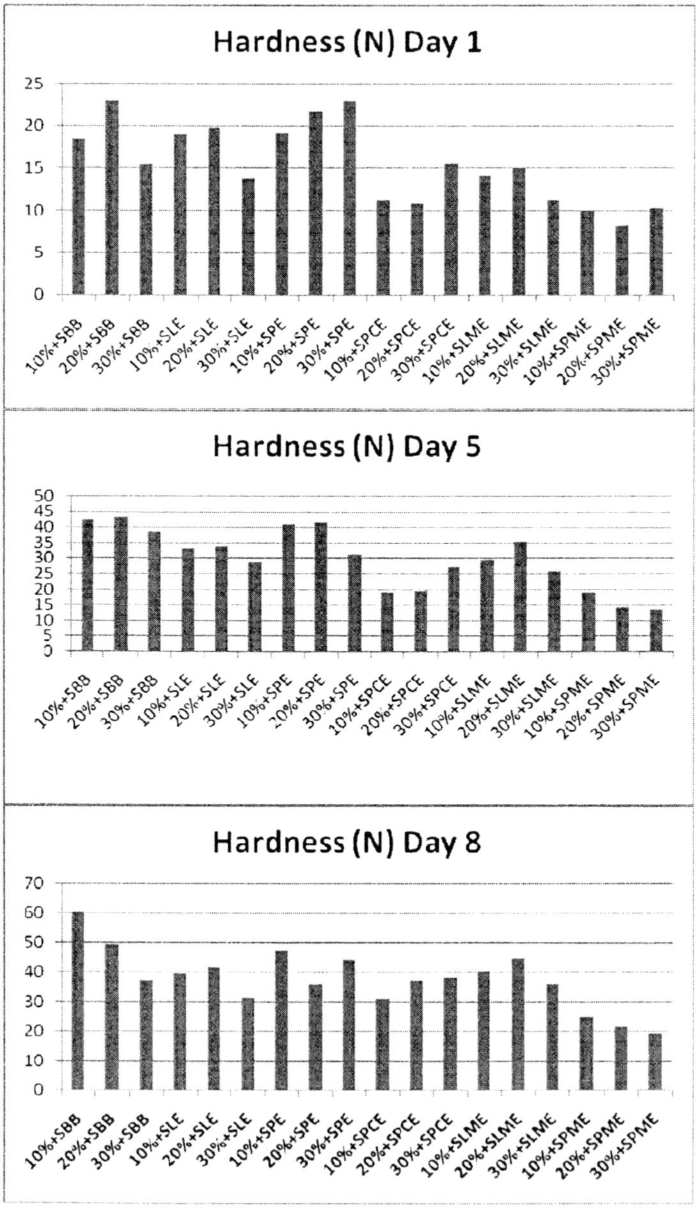

Figure 2 *Maximum hardness of BSG sour dough breads with a range of different enzymes at 1, 5 and 8 days.*

3 CONCLUSION

This study shows that sourdough improves the shelf -life of high-fibre breads compared to conventional dough during 8 days of storage. The combination of enzymes significantly ($P<0.05$) lowered crumb hardness and delayed staling compared to control sour dough and conventional breads while no difference was found in terms of loaf volume.

References

1 V. Stojceska and P. Ainsworth, *Food Chem.*, 2008, **110**, 865.
2 A. Corsetti, B. Gobbetti, B. De Marco, F. Balestrieri, F. Paoletti and J. Rossi, *J. Agric. Food Chem.*, 2000, **48**, 3044.
3 K. Katina, M. Salmenkallio-Marttila, R. Partanen, P. Forssell and K. Autio, *LWT - Food Sci. Technol.*, 2006, **39**, 479.
4 AACC. 1998. Approved method 74-09 of the AACC. St. Paul, Minnesota: American Association of Cereal Chemists.

EFFECT OF PACKAGING CONDITIONS ON SHELF-LIFE OF BOLOGNA SAUSAGES MADE WITH ORANGE JUICE WASTEWATER AND OREGANO ESSENTIAL OIL

M. Viuda-Martos, Y. Ruiz-Navajas, J. Fernández-López, E. Sendra, E. Sayas-Barbera and J.A. Pérez-Álvarez

Grupo Industrialización de Productos de Origen Animal (IPOA), Grupo 1 UMH. Grupo REVIV, Departamento de Tecnología Agroalimentaria, Universidad Miguel Hernández. Crta. Beniel km. 3,2. E-03312 Orihuela, Alicante, Spain.

1 INTRODUCTION

Fruit juices have experienced growing popularity within the last years. In tropical and subtropical areas the predominant fruit juices production and consumption is from citrus fruits and specifically orange. Citrus juice production provides many by-products which constitute about 45–60% of the fruit. Due to the large amount being processing into juice, citrus by-products represent a growing problem for the citrus juice industry.[1] However, their composition has the potential to be used to obtain dietary fibre. The process of obtaining the fibre from citrus fruits also generates a series of by-products, being of special interest the large washing water wastes which, besides having economic impact, are important from an environmental point of view.

One way of avoiding this problem would be to re-cycle the water to take advantage of the large quantity of potentially beneficial substances and to re-use the water, once depurated, which would represent a saving both in economic and environmental terms.

Research is currently being undertaken to improve techniques for extracting the bioactive compounds found in such co-products for use in the production of functional foods.[2] Many of these bioactive compounds are of a polyphenolic nature and are eliminated during the elaboration of foodstuffs. It is therefore important to know exactly which polyphenolic compounds are present in the co-products generated by the food industry in general, and the citrus juice extraction industry in particular. The essential oils of spices can also be used as functional ingredients. Its contain phenolic diterpenes which act as primary antioxidants, showing a high synergistic effect with other antioxidants.[3]
The aim of this work was to study the effect of (i) the addition of orange juice wastewater (OJWW) and oregano essential oil (OEO) and (ii) storage conditions on the chemical, sensory and microbiological properties of bologna sausages.

2 MATERIAL AND METHODS

2.1 Sausage manufacture

Bolognas were manufactured according to a traditional formula (only meat percentages add up to 100% while the percentages of others ingredients are related to meat): 50% lean pork meat, and 50% pork backfat; 15% water (in the form of ice, w/w), 3% potato starch (w/w), 2.5% sodium chloride (w/w), 300 mg/kg sodium tripolyphosphate, 500 mg/kg sodium ascorbate, 150 mg/kg sodium nitrite, spices (0.01% black pepper, 0.005% nutmeg and 0.2% garlic powder). This original mixture was used as control sample.

To assess the influence of the concentration of OJWW, and OEO, water content was replaced in the control formula by OJWW (5%) and OEO were added (0.02%). The orange juice wastewater was obtained by the method described by Fernández-Ginés, et al.[4] and oregano and thyme essential oils were supplied by Ravetllat Aromatics (Barcelona, Spain).

The products were prepared in a pilot plant and followed industrial processing techniques. Frozen raw material of animal origin, except pork backfat, was transferred to the cutter (Tecator 1094 Homogeneizer, Tekator, Höganäs, Sweden) with the sodium chloride to extract salt-soluble proteins; after comminute, the other ingredients and additives were added. Then, pork backfat, previously divided into cubes 10 x 10 x 10 cm, was added. After homogenization, the mixture was stuffed into artificial casing Fibran-Pack (Fibran, Girona, Spain) and cooked in a water bath. The sausages were kept in the bath until the coldest point reached 72 °C. When the endpoint temperature was achieved, the sausages were immediately chilled in ice. After reaching room temperature, the product was transferred to the lab in insulated boxes containing ice.

2.2 Storage conditions

Immediately after manufacture, some slices (1.5 cm thick) were aseptically removed from each bologna sausage. The trays containing the slices were packed either in vacuum, and air pouches made of polyethylene and polyamide laminate of water vapour permeability 1.1 $g/m^2/24$ h at 23 °C, nitrogen permeability 10 $cm^3/m^2/24$ h at 23 °C, carbon dioxide permeability 140 $cm^3/m^2/24$ h at 23 °C, and oxygen permeability 30 $cm^3/m^2/24$ h at 23 °C (Fibran, Girona, Spain). The pouches were heat-sealed and stored at 4 ± 1 °C in a cabinet simulating supermarket retail conditions. The cabinet was illuminated by a standard supermarket fluorescent lamp (OSRAM, Germany). All the samples were exposed to continuous lighting at 1000 lux at the surface. The positions of the samples in the cabinet were rotated every 24 h to minimize light intensity differences and possible temperature variations on the meat surface. The packs were stored for 24 d. Samples from each treatment and storage conditions were taken at 0, 6, 12, 18, and 24 d (storage time) and analyzed on the same day.

2.3 Lipid oxidation

Lipid oxidation was assessed in triplicate by the 2-thiobarbituric acid (TBA) assays following the recommendations of Buege and Aust.[5] TBARS values were calculated from a standard curve of malonaldehyde (MAD) and expressed as mg MAD/kg sample.

2.4 Sensory evaluation

Non-trained panellists (30) were recruited from the staff and students of the Miguel Hernández University, Alicante, Spain. Panellists were chosen on the basis of previous experience in consuming traditional bolognas. Furthermore, a preparatory session was held prior to testing, so that each panel could thoroughly discuss and clarify each attribute to be evaluated in bolognas. Testing was initiated after the panellists agreed on the specifications. A Quantitative Descriptive Analysis was carried out.[6] During evaluation, the panellists set in private booths under incandescent/fluorescent light, with an intensity of approximately 350 lux. Rectangular pieces approximately 1.5 cm × 2 cm were cut from the centre of bologna slices and served at room temperature. Each panellist evaluated three replicates of all the treatment groups; the sample presentation order was randomized for each panellist. Tap water was provided between samples to cleanse the palate. The sensory attributes were measured in unstructured scales with descriptors at both ends, no standards were provided. The attributes measured and their descriptors were as follows: for "external evaluation": global appearance (from conventional bologna appearance to unexpected appearance), colour (from extremely light to extremely dark), shine (from dull to bright), hue (from pale pink to brown) and homogeneity perception (from particulate to no particles observed); for odour (from imperceptible to extremely intense); for "taste": acid taste, saltiness, and fatness (from imperceptible to extremely intense); and for "texture": hardness (from extremely soft to extremely tough), juiciness (from extremely dry to extremely moist). At the end of the test, panellists were asked to give a score for product quality from 0 to 10.

2.5 Microbiological analysis

Samples (25 g) were excised from the interior of the sausages with a sterile scalpel and forceps and then homogenized with sterile 1.5% peptone water in a Stomacher 400 (Colworth, London, U.K.) for 1.5 min. Total viable counts were determined on Plate Count Agar, *Enterobacteriaceae* using Violet Red Bile Glucose Agar (VRBGA) and Lactic acid bacteria (LAB) were counted on double layer MRS Agar at pH 5.6. In all cases, plates were incubated at 35 °C for 48 h. Psychrotrophic microbiota was determined on Plate Count Agar, and the plates were incubated at 7 °C for 10 d. Culture media were from Oxoid (Oxoid Basingstoke, Hampshire, U.K.).

2.6 Statistical analysis

Conventional statistical methods were used to calculate means and standard deviations. Statistical analysis (ANOVA) was applied to the data to determine differences ($P < 0.05$). To discover whether there were significant differences between the levels of the main factor, contrasts (Tukey test) between means were made.[7] For the bologna self-life determination, ANOVAs with 3 factors (storage time: 0, 6, 12, 18 and 24; packaging conditions: vacuum and air; treatments: control and OJWW+OEO) were applied for each parameter. The Statistical analyses were made using Statgraphics 5.1 for Windows.

Table 1 *TBA values (mg MAD/kg sample) of bologna control and bologna formulate with OJWW+OEO, stored on vacuum package and air package, during 24 days of storage.*

Sample	Storage conditions	Time (d) 0	6	12	18	24
Control	Air	4.15 ± 0.03^{aw}	5.41 ± 0.01^{bw}	6.01 ± 0.01^{cw}	6.52 ± 0.06^{dw}	7.14 ± 0.00^{ew}
	Vacuum	4.15 ± 0.03^{aw}	5.34 ± 0.00^{bx}	5.76 ± 0.02^{cx}	6.18 ± 0.04^{dx}	6.26 ± 0.00^{ex}
OJWW+OEO	Air	4.10 ± 0.00^{az}	5.39 ± 0.01^{bw}	5.98 ± 0.01^{cw}	6.40 ± 0.00^{dy}	6.94 ± 0.01^{ey}
	Vacuum	4.10 ± 0.00^{az}	5.28 ± 0.01^{by}	5.60 ± 0.02^{cy}	5.97 ± 0.03^{dz}	6.15 ± 0.00^{ez}

$^{a-e}$ Means within a row with different letters are significantly different ($p< 0.05$)
$^{w-z}$ Means within a column with different letters are significantly different ($p< 0.05$)

3 RESULTS AND DISCUSSION

3.1 Lipid oxidation

Chopping is one of the most important steps in meat emulsions manufacturing. Chopping and heating can catalyze the lipid oxidation because they disrupt cellular protective compounds contained in cell membranes such as vitamin E and hydrogen donors.[8] Lipid oxidation was evaluated by determining the levels of TBARS (Table 1).

At day 0, the OJWW+OEO samples showed lower oxidation values ($P<0.05$) than the control. At day 6, the lowest oxidation values recorded were in the OJWW+OEO samples packed in vacuum while no statistically significant differences were observed between the air packed control sample and the OJWW+OEO sample air packed. At days 12 and 18, the vacuum-packed OJWW+OEO samples continued to show the lowest oxidation values, while the air-packed OJWW+OEO sample showed higher ($P<0.05$) values than the vacuum-packed control (6.40 and 6.18 mg MAD/kg sample, respectively). At the end of the experiment (24 days), the vacuum-packed OJWW+OEO treated sample showed the lowest oxidation degree ($P<0.05$) of all the samples, regardless of packaging type, while both the air-packed control and air-packed OJWW+OEO samples showed the highest values for this parameter.

The antioxidant activity of co-products from the industrial manipulation of oranges has been widely demonstrated.[4,9] Such activity is basically due to their composition: mainly due to phenolic compounds and flavonoids. However, the action mechanism responsible for the antioxidant activity is not clear, although several mechanisms have been proposed such as the scavengers of free radicals, hydrogen donation, metallic ion chelation or even acting as substrate of radicals such as superoxide or hydroxyl.[10] The activities of flavonoids such as antioxidants depend not only on their structural features but also on many other factors, such as concentration, temperature, light, type of substrate, physical state of the system, as well as on microcomponents acting as pro-oxidants or synergists.[11] Furthermore, components of essential oils, including thyme and oregano EOs, show their own antioxidant activity.[12,13] These bioactive compounds with antioxidant activities may also interfere with the propagation reactions[14] besides inhibiting the enzymatic systems involved in initiation reactions.[15]

3.2 Sensory analysis

Figure 1 shows the results obtained for the sensory evaluation carried out at the end (day 24) of the assay. The panellists clearly detected the loss of colour produced in the samples. The fat character, acid taste, salty taste and odour values were very similar. It should be noted that despite the marked aroma of oregano, this was not found unpleasant by the panellists, who valued the samples containing this essential oil almost equally with the controls. In general, greatest differences for all parameters were detected between control and OJWW+OEO samples, control samples obtained the poorest scores.

The best scored sample was the OJWW+OEO sample packaged in vacuum, and the least acceptable the control sample exposed to air.

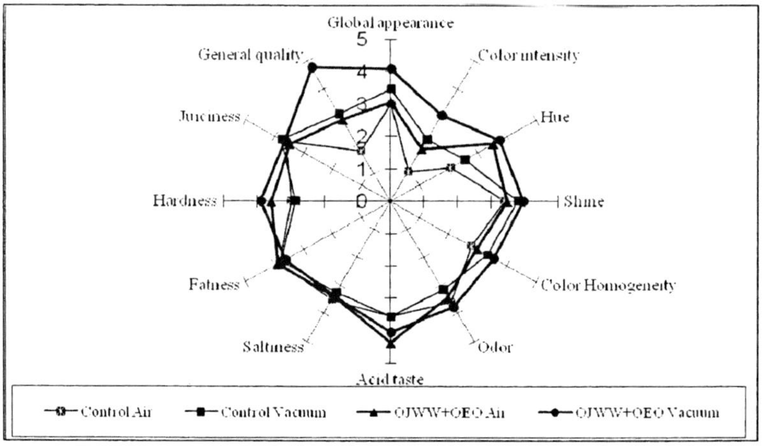

Figure 1 *Quantitative descriptive analysis carried out at day 24 of control bolognas and bolognas formulated with Orange juice waste water and oregano essential oil.*

3.3 Microbiological analysis

The growth of aerobic and lactic bacteria was greater (P<0.05) in the control samples than in those with OJWW+OEO (Figure 2 a and b) at all times and in all packaging conditions,

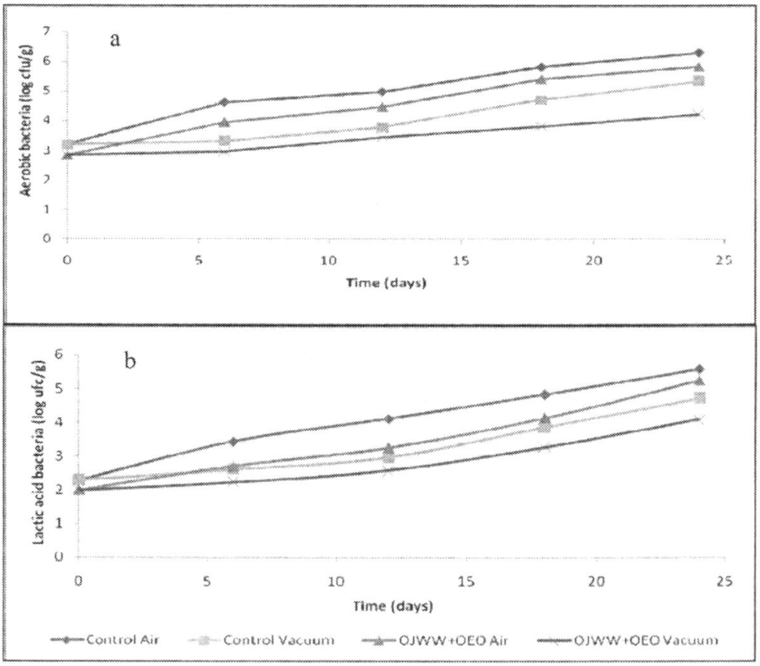

Figure 2 *Evolution of counts of aerobic bacteria (a) and counts of lactic acid bacteria (b) of bolognas control and formulated with orange juice waste water and oregano essential oil.*

except in the OJWW +OEO sample packed in air at six days, which showed higher values than the controls packed in vacuum, a difference that was maintained until the end of the experiment (24 days). No *enterobacteria* or psycotrophic bacteria were found in any of the treatments (control and OJWW+OEO), regardless packaging and storage time of storage, probably due to the effectiveness of the cooking, aseptic slicing process together with the presence of the sodium chloride contained in the products.[16]

In all the samples, the total aerobic bacteria and lactic acid counts at the end of the experiment were below those considered as representing a degraded product (higher viscosity, colour changes, off-flavours).

4 CONCLUSIONS

The addition of orange juice wastewater and oregano essential oil seems to be a viable alternative for elaborating cooked meat products because of the "natural" image of the products is improved. Their addition increase the acceptance rates of the products their oxidative stability and the shelf-life of the products and reduce microbial growth.

References

1 J. Fernández-López, E. Sendra-Nadal, C. Navarro, E. Sayas, M. Viuda-Martos and J.A. Pérez-Alvarez, *Int. J. Food Sci. Tech.* 2009, **44**, 748.
2 M. Viuda-Martos, J. Fernández-López, E. Sayas, J.A. Pérez-Alvarez and E. Sendra-Nadal, *Alimentación, Equipos y Tecnología* 2007, **221**, 64.
3 M. Viuda-Martos, J. Fernández-López and J.A. Pérez-Alvarez, *Alimentación, Equipos y Tecnología*, **229**, 40.
4 J.M. Fernández-Ginés, J. Fernández-López, E. Sayas-Barberá, E. Sendra and J.A. Pérez-Alvarez, *J. Food Sci.* 2003, **68**, 710.
5 J.A. Buege and S.D. Aust, *Meth. Enzymol.* 1978, **52**, 302.
6 IFT, *J. Food Sci.* 1981, **11**, 50.
7 A.A. Afifi and S.P. Azen, *A computer oriented approach,* London Academic, 1978, p. 18.
8 T. Keokamnerd, J.C. Acton, I.Y. Han and P.L. Dawson, 2008, *Poultry Sci.* **87,** 170.
9 J. Fernández-López, M. Viuda-Martos, E. Sendra, E. Sayas-Barberá, C. Navarro and J.A. Pérez-Alvarez, *Eur. Food Res. Tech.* 2007, **226,** 1.
10 M. Al-Mamary, A. Al-Meeri and M. Al-Habori, *Nutr. Res.* **22**, 1041.
11 N.V. Yanishlieva-Maslarova, *Antioxidants in food: Practical applications*, Woodhead Publishing Ltd. 2001, p 22.
12 K.A. Youdim, S.G. Deans and H.J. Finlayson, *J. Essen. Oil Res.* 2002, **14,** 210.
13 T. Kulisic, A. Radonic, V. Katalinic and M. Milos, *Food Chem.* 2004, **85,** 633.
14 A. Russo, R. Aquaviva, A. Campisi, V. Sorrenti, C. Di-Giacomo, G. Virgata, M.L. Barcellona and A. Vanella, *Cell Biol. Toxicol.* 2000, **16**, 91.
15 K.M. You, H.G. Jong and H.P. Kim, *Arch. Pharm. Res.* 1999, **22,** 18.
16 E. Borch, M.L. Kant-Muermans and Y. Blixt, *Int. J Food Microbiol.* 1996, **33,** 103.

EFFECTS OF INDUSTRIAL PROCESSING ON CONTENT AND PROPERTIES OF DIETARY FIBRE OF STRAWBERRY WASTES

P. Torres,[1] F.J. López-Andréu,[2] G. Torres,[2] M. Vidriales,[2] R.M. Esteban,[2] E. Mollá[2] and M.A. Martín-Cabrejas[2]

[1]Hudisa Desarrollo Industrial S.A., "La Gravera" Apdo. 46, 21440 Lepe (Huelva), Spain.
[2]Dpto. Química Agrícola, Universidad Autónoma de Madrid, 28049 Madrid, Spain. (maria.martin@uam.es)

1 INTRODUCTION

Strawberry is a well known fruit of frequent human consumption. In Europe the main producer of strawberry is Spain, mainly the south west of the country, being Huelva the principal area of production due to its own specific conditions of soil, climate and availability of good quality water. The annual production is 312,000T of strawberry per year that represents 94% of the national production and generates 420 10^6 € annually. *Fragaria ananassa* Var. California camarosa is the most cultivated and competitive variety in the European market for several reasons: production period, high productivity, sensory quality and adaptation to soil and climatic conditions.

Color, texture, odor and the balance between sweetness and sourness have been identified as important quality factors of strawberry fruit. Previous studies determined the major components of strawberry, during development and ripening, which are responsible of quality attributes.[1] Phenolics, carotenoids, organic acids, sugars, polysaccharides such as pectic substances and other compounds of dietary fiber (DF) and ascorbic acid are the main components in the strawberry composition.

The 90% of strawberry production is derived to fresh consumption and half of it is exported to different European countries. The rest 10% is considered surplus that in most of the cases is processing (≈ 20.000 T). Thus, there are high levels of surplus, constituted by strawberries discarded during the selection and calibration steps. These wastes represent a high percentage of weight from raw material. The strawberry producers are interesting in finding solutions to reduce the high level of surplus from strawberry, searching for ways of utilization to obtain new and natural sources of high value functional ingredients and to guarantee the cost-effectiveness of the agri-food exploitation in this area before depressed. Low cost industrial processes should be an adequate way to convert these strawberry wastes into useful products. The production of fibre-enriched foods by incorporating DF sources such as fruits would increase the DF intake and thus to reach the recommended (25-35 g per day) level, as well DF from fruit exhibits a balanced ratio between soluble and insoluble fractions.

The objective of the present work was to provide precise information of the contents and properties of dietary fibre of strawberry intermediate products from industrial processing in order to evaluate their potential use as sources of dietary fibre.

2 MATERIAL AND METHODS

2.1. Material

Fresh strawberries were washed, sieved and homogenized, and enzymatic inactivation was carried out, obtaining the first product named Pulp. The second product Purée was obtained after grinding and sifting the pulp, followed by pasteurization. Finally a Concentrated product was obtained after vacuum evaporating the water. These processed products and fresh fruits were frozen at -80 °C and lyophilized at -45 °C and vacuum for 72 hours, samples were ground and passed through a 0.5 mm sieve and stored at -20 °C until analysis.

2.2. Methods

2.2.1. Moisture content was determined according to standard method.[2]

2.2.2. Dietary Fibre (DF). Mes-Tris AOAC method 991.43 was used for DF determination;[2] the insoluble residues were isolated by filtration and soluble fiber was precipitated with ethanol. Dried residues correspond to insoluble dietary fiber (IDF) and soluble dietary fiber (SDF), respectively. Determination of residual ashes and proteins (as Kjeldahl N x 6.25) was carried out in the residues, according to standard procedures[2] for corresponding corrections. Total dietary fiber (TDF) was calculated as sum of IDF and SDF.

2.2.3. Chemical Composition of cell wall. Chemical composition was determined from the AIR (Alcohol Insoluble Residue) obtained according to Yeh *et al.*[3] The neutral sugar composition of the DF was determined by HPLC[4]. Uronic acids were determined colorimetrically.[5]

2.2.4. Physicochemical Properties of the Fibre. They were determined on AIR.
Bulk Density: It was determined according to Singh,[6] and calculated as grams per millilitre.
Water Holding Capacity (WHC): According to method of Chau and Huang,[7] with slight modifications. The WHC was expressed as mL of water held by one gram of dry sample.
Swelling Capacity: expressed as volume (mL) occupied by sample per gram of dry original sample weight.[8]
Oil Holding Capacity (OHC): According to method of Chau and Huang,[7] with slight modifications. The OHC was expressed as mL of vegetal oil held by one gram of dry sample.
Least Gelation Concentration: The gelification activity was determined as the smallest concentration in which the samples do not slide by tube when it is inverted.[9]

3 RESULTS AND DISCUSSION

The intermediate products Pulp and Purée present high moisture, similar to fresh strawberry (93%), while in case of Concentrated product decreased to 85%. Regarding to DF (Figure 1) the intermediate products exhibited a decrease of TDF content, mainly due to IDF fraction. Either the losses in sieving residues or the thermal processing effects promoted a decrease of fibre. The insoluble: soluble ratio of Pulp, Purée and Concentrated was 1:1. Consequently, the intake of DF from strawberry products may have beneficial physiological effects due to both insoluble and soluble fractions; therefore strawberry wastes appeared to have a potential use a DF source.

The sugars composition of cell wall of fresh fruit and intermediate products are showed in Figure 2. The strawberry is rich in pectin, and intermediate products showed great solubilization of pectic substances. There is an increasing interest in pectins for their potential to bring down blood cholesterol levels, by decreasing the low-density-lipoprotein-cholesterol fraction without changing the levels of high-density-lipoprotein-cholesterol and triglycerides. Pectins also affect glucose metabolism by lowering the glucose response curve.[10] Besides clinical applications, pectin is used as a food additive because of its specific properties as a gelling agent.

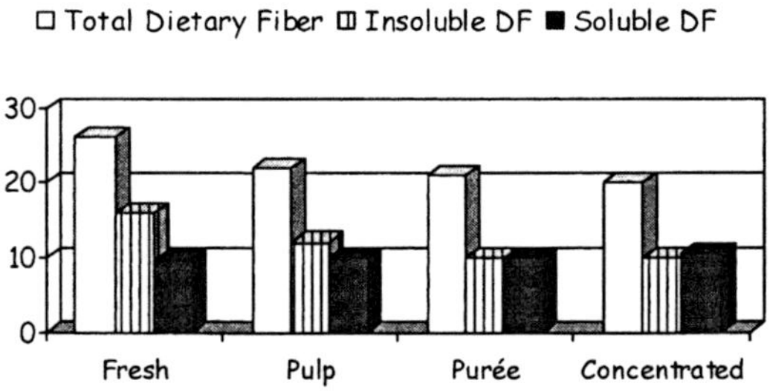

Figure 1 *Total, Insoluble and Soluble Dietary Fibre (g/100 g dry matter)*

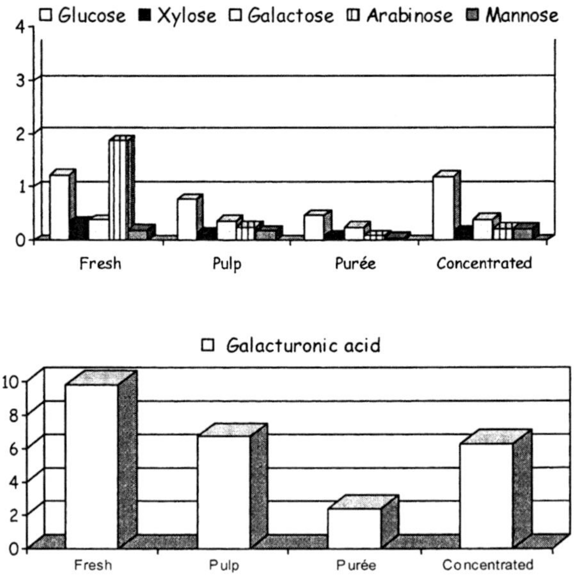

Figure 2 *Sugar Composition of cell wall (g/100 g dry matter)*

The industrial processing can change physico-chemical characteristics of DF, due to the variations of their components (Figure 3, 4, 5, 6 and 7). Regarding to Bulk density of these dietary-rich materials (Figure 3), the Purée was significantly ($p<0.05$) more dense than the Concentrated product, being both more dense than fresh strawberry. Other studies about fruit and vegetable products showed higher Bulk density values than these strawberry derivatives.[7,11] In any case, all intermediate products were less dense than cellulose; it means these materials exhibited higher porosity which will be related to water holding capacity (WHC). This property depends on the structural characteristics typical of each material, the particle size and their distribution.

Regarding to WHC (Figure 4), all materials showed higher WHC than cellulose and than other vegetal products;[3,11,12] fresh strawberry and pulp exhibited greater level than Pureé and Concentrated, which could be attributed to a lower IDF content in these latter products. The effect of processing was significant, showing a drastic decrease of WHC in both products, which showed similar levels (15-14 mL/g). Nevertheless, the high WHC of these materials compared to other plant products suggested that they could be used as a functional ingredient to avoid syneresis and to modify the viscosity and texture of formulated products in addition to reducing calories by the total or partial substitution of high-energy ingredients.[13] The WHC is desirable in food such as sausages, custards and doughs because they are supposed to imbibe water thereby attaining body thickening and viscosity.

In addition, the hydrophilic tendency was also measured by Swelling Capacity (Figure 5). The results obtained exhibited the same trend than WHC and always with lesser WHC values than Swelling Capacity values; similar results were obtained for other studies.[14] As compared to the control cellulose the higher swelling abilities obviously mean that they have higher porosity and larger surface area, and the higher porosity is in agreement with their lower Bulk density.

In relation to Oil Holding Capacity (OHC) (Figure 6), the studied intermediate products exhibited similar values, from 1.7 for fresh to 2.1 mL/g for pulp, and the lipophilic tendency was lesser than cellulose. OHC of the samples did not show any changes by thermal processing. The OHC values of the samples were higher than orange DF concentrates,[10] pea fibre,[15] and cauliflower fibre.[16] Not much information was found about the OHC of DF from other fruits and agricultural by-products.

Gelation capacity, essential property in the preparation and acceptability of many foods, was studied using the Least Gelation Concentration (LGC) as the index of gelation capacity, therefore low LGC means better gelation property. The Least Gelation Capacity of strawberry materials and cellulose was similar (Figure 7).

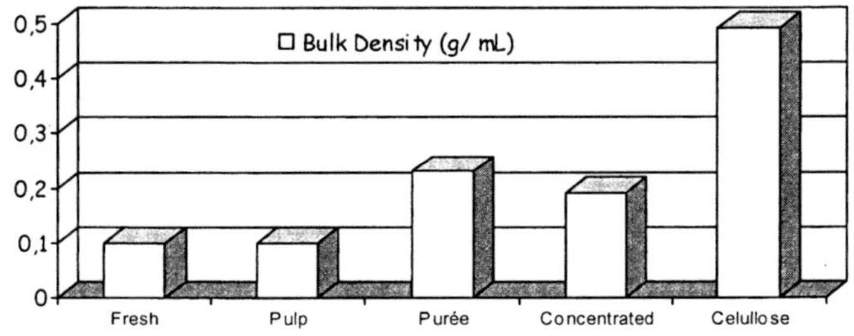

Figure 3 *Physico-chemical properties of DF. Bulk density*

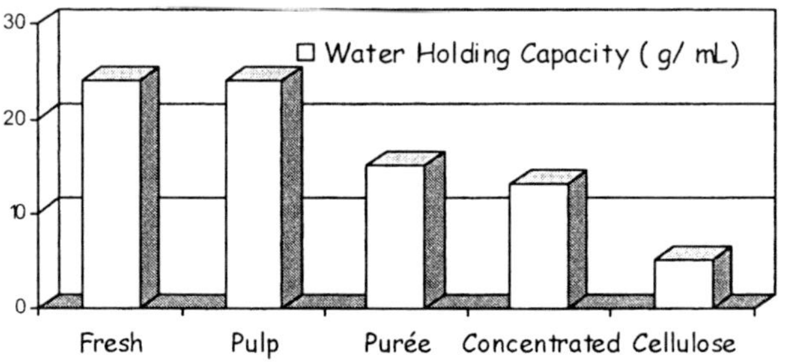

Figure 4 *Physico-chemical properties of DF. Water Holding Capacity (WHC).*

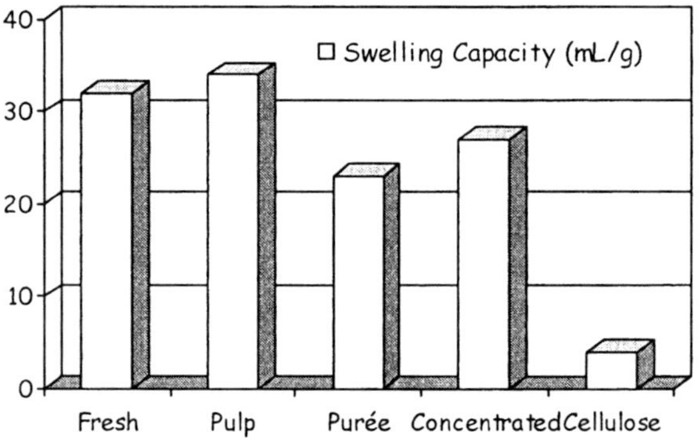

Figure 5 *Physico-chemical properties of DF. Swelling Capacity.*

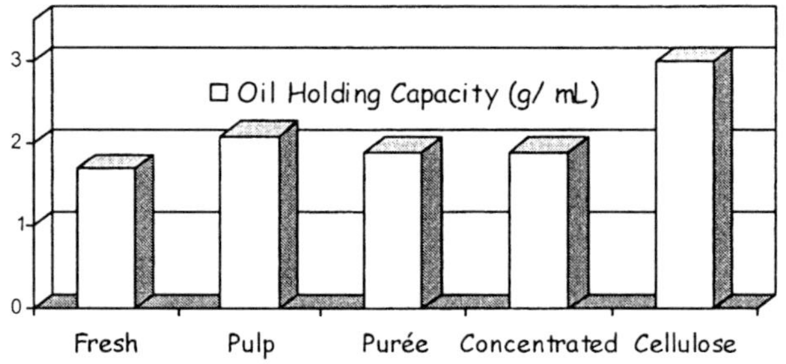

Figure 6 *Physico-chemical properties of DF. Oil Holding Capacity (OHC).*

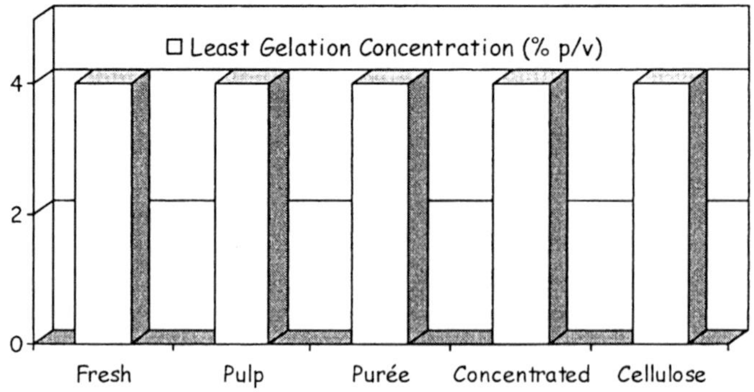

Figure 7 *Physico-chemical properties of DF. Least Gelation Concentration (LGC)*

4 CONCLUSIONS

From the results obtained in this study, it can be concluded that strawberry intermediate products turned out to be an adequate source of DF (20-22% d.m.), since industrial processing caused only slight decreases of TDF and IDF. Related to functional properties, it has been shown, the high WHC and OHC of these materials that might be attributed to their lower bulk density and high swelling ability. Therefore they could use as ingredients in food applications requiring moisture and oil retention. These intermediate products appeared to be versatile ingredients that combined a natural origin, a well-balanced fibre content, great functional properties and a low energy value for use by food industry.

References

1 T.M. Montero, E. Mollá, R.M. Esteban and F.J. López Andréu, *Sci. Hort.,* 1996, **65**, 239.
2 AOAC. *Official Methods of analysis*,16th ed., Washington, D.C. 1995.
3 H-Y. Yeh and N-W. Su, *J. Agric. Food Chem.* 2005, **53**, 4361.
4 L. Jaime, E. Mollá, A. Fernández, M. Martín-Cabrejas, F-J. López-Andréu and R.M. Esteban, *J. Agric. Food Chem.*, 2002, **50**, 122.
5 R.M. Esteban, F.J. Lopez-Andréu, M.A. Martin-Cabrejas and E. Mollá, *Food Chem.* 1993, **46**, 289.
6 U. Sing, *J. Food Sci. Technol.*, 2001, 38, 191.
7 C.F. Chau and Y.L Huang, *J. Agric. Food Chem*, 2003, **51**, 2615.
8 G.L. Robertson, F.D. Monredon, P. Dysseler, F. Guillon, R. Amado and J.F. Thibault, *LWT Food Sci. Technol.*, 2000, **33**, 72.
9 C.F. Chau and P.C.K. Cheung, *Food Chem.*, 1997, **61**(4), 429.
10 N. Grigelmo-Miguel and O. Martin-Belloso, *Food Res. Int.*, 1999, **31**, 355.
11 C.F. Chau, C.H. Chen and M.H. Lee, *LWT Food Sci. Technol.*, 2004, **37**, 155.
12 C.F. Chau and Y.L. Huang, *Food Chem.* 2004, **85**, 189.
13 N. Grigelmo-Miguel, S. Gorinstein and O. Martin Belloso, *Food Chem.* 1999, **65**, 175.
14 A. Femenia, P. García-Pascual, S. Simal and C. Roselló, *Carbohydr. Polym.*, 2003, **51**, 397.
15 M.-C. Ralet, G. Della Valle and J.-F. Thibault, *Carbohydr. Polym.*, 1993, **20**, 17.
16 A. Femenia, R.R. Selvendran, S.G. Ring and J.A. Robertson, *J. Agric. Food Chem.*, 1997, **47**, 728.

FORMULATION AND ACCEPTABILITY STUDIES OF HIGH FIBER COOKIES MADE FROM PINK GUAVA (*PSIDIUM GUAJAVA*) DECANTER / AGRO WASTE

H. Chek Zaini, H. Zaiton, C.W. Zanariah and N. Sakinah

Food Biotechnology, Faculty Science and Technology, University Sains Islam, Malaysia.

1 INTRODUCTION

Guava fruits are rich in vitamins A and C and contains high amount of dietary fiber.[1-3] The fruits are usually large in size, crunchy like apple, harvested while the peel is still green, and is eaten raw. They are marketed mainly through the Asian specialty grocery stores and farmer's markets. Tropical guavas mainly consumed by Hispanics are picked at a slightly more matured age showing some colours of the peel, and usually are eaten when slightly soft. Guava fruits are undesirable when it is too ripe, as it has a mushy texture and very strong odour. In Malaysia, guava fruits are harvested in abundance, but short lives, rots and decays easily when ripe; resulting in wastage. The progression of time, with the advancement of technology and the desire to solve this problem has become a platform for researches local and abroad to seek alternative solution in varying the production of products based on fruits. Golden Hope Food & Beverages Sdn. Bhd. Malaysia is the largest pink guava producer in Asia with over 500 hectares guava planted that produced 10,000 tons of fruits for pink guava juice for exports to Japan, USA, Australia, Philippines, Korea, Canada, Singapore and New Zealand. From the 10,000 tons of fruits, about 10% (100 tons) of the fruits are considered decanter waste (scrubs and seeds). With this facts in mind, and the loves of guava fruits taste, triggered the idea to study the possibility of using different components of guava decanter waste in producing high fiber products such as cookies and flakes. Every 100gm of scrubs of the decanter waste contain 120 mg ascorbic acid and the guava seeds contain 145 of aromatic oil, 15% protein, 36% fiber contents (encompassed of cellulose and lignin) and 13% starch. The guava leaves are rich in tannin (10% in the leaves on a dry weight basis). The leaves are used to employ for gastroenteritis, diarrhea and dysentery. Crushed leaves are applied on wounds, ulcers and rheumatic places and also chewed to relieve toothache, remedy for coughs, throat and chest ailments, oral ulcers and inflamed gums, diarrhea, nephritis and cachexia. In Japan, guava leaves are also used to produce guava tea.[7]

2 MATERIALS AND METHOD

2.1 Preparation of Cookies Variation

Cookies are prepared using butter, sugar, salt, eggs, vanilla, commercial fine flour and baking powder, and it will be processed by creaming, rolling and baking at temperature between 350-375 °F for 12 minutes. Mixture tested for the study using variations of 5, 10, 15, 20, 25, 30 and 40% were prepared by substituting cake flour with guava solids, scrubs, seeds and combination of sieved seeds and leaves. The guava solids were obtained from Golden Hope Food & Beverage Sdn. Bhd., Sitiawan, Perak.

Cookies were prepared using the following ingredients (Table 1). All the ingredients were purchased at a local supermarket.

Table 1 *Basic cookie ingredients for control.*

Ingredients	U.S		Metric		%
Butter	3.1	oz	88	g	40
Sugar	4	oz	110	g	50
Salt	0.07	oz	2	g	0.9
Eggs	0.78	oz	22	g	10
Vanilla extract	0.1	oz	3	ml	1.5
Commercial fine flour	7.8	oz	220	g	100
Baking Powder	0.16	oz	4.4	g	2
Total weight:	***1 lb 1***	***oz***	***449.4***	***g***	***204.4%***

Mixing: Creaming method.
Make up: Rolled method. Greased or parchment lined baking sheets used.
Baking: 375 °F (190 °C) for 10-12 minutes.

Blends of 5, 10, 15, 20, 25, 30, 35 and 40% were prepared by substituting cake flour with guava solids: scrubs, seeds and combination of sieved seeds and leaves. Guava seeds grounded. The dried leaves are crushed using a mortar and pestle. Test baking of the cookie was carried out for blends containing different levels of guava solids according to the methods of Ajila *et. al.*,[4] Gisslen[5] and Leelavathi & Haridas Rao.[6] The formula used was as follows: 220 g flour, 110 g sugar, 88 g butter, 4.4 g baking powder, 22 g eggs, 3ml vanilla extract and 2 g salt. Butter, sugar and salt were creamed in a mixer (KitchenAid, Belgium) with a flat beater for min at 61 rpm (speed 1). Eggs and vanilla extract were added to the cream. The contents were mixed for 6 min at 125 rpm (speed 2) to obtain a homogeneous cream. Sieved flour and baking powder was added to the above cream and hand-mixed with a rubber spatula until just combined. Cookie dough was sheeted to a thickness of 3.5 mm, cut using a circular mould (40 mm diameter) and baked in an electric oven (Tecnoeka, Italy) at 375 °F (190 °C) for 10-12 minutes. After baking, cookies were cooled to room temperature and were packed in polypropylene pouches and sealed until further analysis.

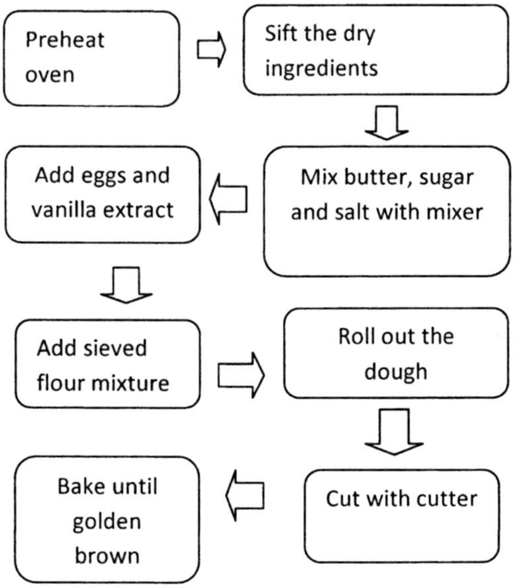

Figure 1 *Flow process of guava cookies*

3 RESULTS AND DISCUSSIONS

3.1 Evaluation of Cookies

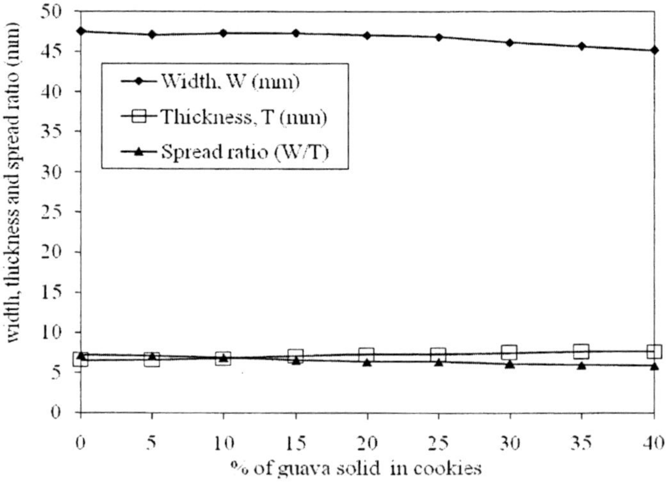

% of guava solid	0	5	10	15	20	25	30	35	40
Width (mm) ◆	47.52	47.12	47.33	47.33	47.05	46.87	46.14	45.67	45.17
Thickness (mm) □	6.55	6.6	6.83	7.15	7.33	7.33	7.5	7.63	7.65
Spread ratio (mm) ▲	7.25	7.14	6.93	6.62	6.42	6.39	6.15	5.99	5.90

Figure 2 *Trend line of % guava solid versus width, thickness and spread ratio of cookies*

Physical characteristics of the cookies were evaluated and inserted into several line graphs shown in Figure 2 to study the effects of incorporating guava solids. The trend line of the width (mm) of the cookies decreases with the increasing amount of guava solids incorporated. Whereas the trend line of the thickness of the cookies increases with the amount of solids incorporated. Decrease in the width or diameter with the addition of guava solids may be due to the dilution of gluten,[4] the thickness of the cookies increases with the increasing amount of guava solids incorporated may be due to the nature of guava solids, which are not as fine as the commercial flour used. With the increase of guava solids in the cookies, the width of the cookies slightly decreases, while the thickness increases. Although it may not seem significant by the naked eye, calculation and thus the average taken results in the decreasing spread ratio. However, according to Kirssel & Prentice,[15] cookies having higher spread ratios are considered desirable.

Nevertheless, other researchers also reported the reduction in spread ratio when oat bran, soy flour and black gram flour were substituted for wheat flour.[12-14,17] Hooda & Jood[10] reported the reduction of spread ratios with fenugreek flour substitution, while Sudha *et al.*[9] reported reduction of spread ratio with wheat, rice, oat and barley bran. Reduced spread ratios of these cookies may be due to the fact that composite flour apparently form aggregates with increased numbers of hydrophilic sites available for competing for the limited free water in cookie dough.[10,16]

3.2 Crude Fiber Content

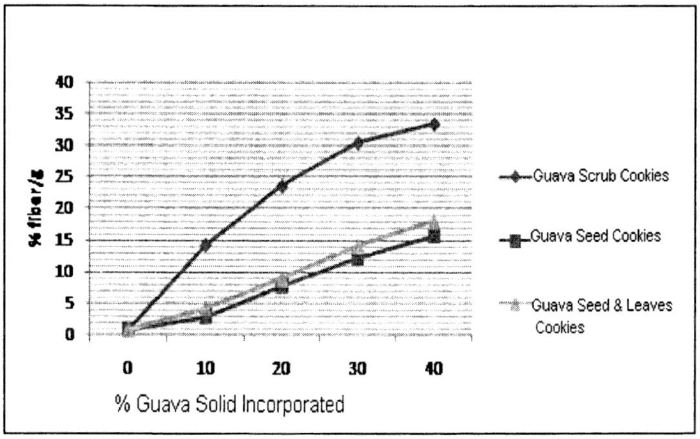

Figure 3 *% Fiber Average for Cookie Variation*

Figure 3 shows the percentage of fiber content in cookies increases with the addition of guava solids. Guava scrub cookies contain the highest amount of fiber, followed by guava seeds and leaves cookies, and guava seed cookies. This also shows that the addition of crushed leaves scattered on the cookies, do have a difference in the percentage of fiber resulted. When ANOVA was carried out to test the significant difference between the control cookie with no guava solid incorporated, all samples showed significantly different values. The test statistic, $F = 1018.72$ for guava scrub cookies; $F = 964.69$ for guava seed cookies and $F = 855.17$ for guava seeds and leaves cookies; with all having a p-value of

0.00. The p-value is the probability of obtaining a test statistic as large as those F values obtained, assuming H_0 is true. Since the p-value is less than $\alpha = 0.05$, the null hypothesis is rejected. We have sufficient evidence to conclude that the true mean control differ for at least one of the mean from the sample cookie variation.

3. 3 Protein Content

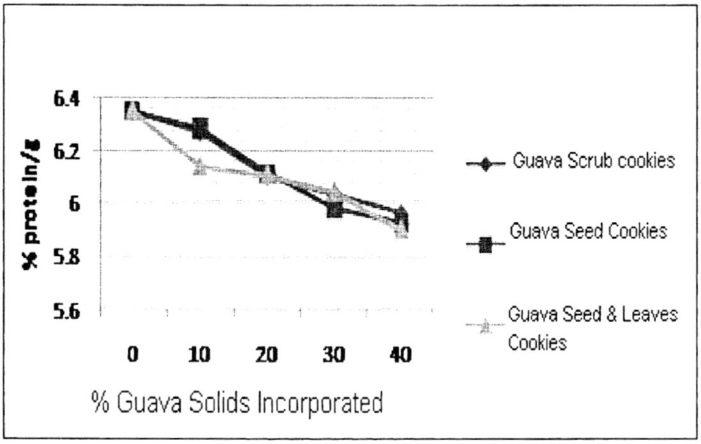

Figure 4 *% protein average for cookie variation*

Figure 4 shows the addition of guava solids in cookies results in a decrease of protein, compared to the control. The average % protein of control cookie of 0% guava solids are 6.35% and decreases to 5.96%, 5.92% and 5.9% for 40% guava scrubs, seeds and seeds and leaves cookies respectively. Similar behavior was reported by Rathi *et al.*[5] where decreases of protein occur with the higher incorporation of de-pigmented pearl millet in biscuits. This suggests that the lower values of protein may be attributed partly to the hydrolysis and partly to the leaching of protein. This is also due to the naturally amount of commercially fine flour in the formulations.

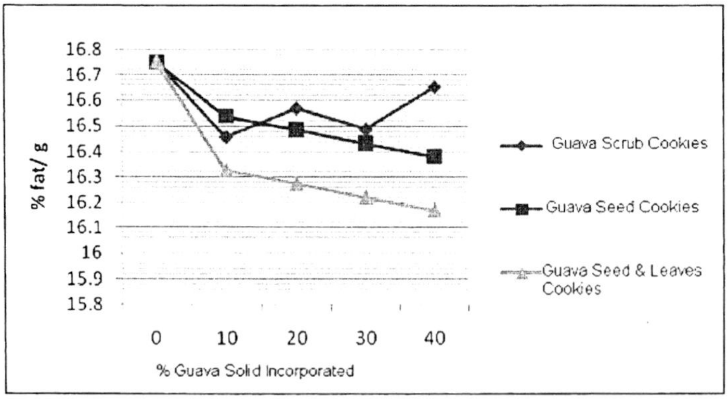

Figure 5 *% fat average for cookie variation*

3.4 Fat Content

Figure 5 shows the trend line of fat is inconsistence over % guava solids incorporation in cookies with the value of fat ranges only from 16.2-16.8%. There is only a 0.6% difference between the highest and the lowest value. This shows that the decrease amount of protein to the decrease percentage of guava solids does not influence the percentage of fat significantly ($\alpha = 0.05$) due to the fact that the amount of fat (butter) is kept constant in all formulations.

3.5 Texture Analysis

The mastication energy went up from 0.72 up to 7.02 N mm (Figure 6) with the incorporation of guava scrub cookie, indicating that guava solids do influence the hardness of cookies. The masticable quality is the multiplication value of hardness with elastic quality and cohesive quality; measured using the texture analyzer (Shimadzu, Japan), where:

Masticable quality@ mastication energy = hardness x elastic quality x cohesive quality

This means that the higher amount of guava solids used in cookies, the harder the cookies will be. The increase in hardness in cookies may be due to the relatively higher water content in guava solids incorporated doughs. This is because, according to Ajila *et al.*,[2] doughs having higher water content produce an extensive gluten structure and result in harder cookies. Recently, Sudha *et al.*[3] also reported an increase of breaking strength of cookies upon the addition of cereal bran.

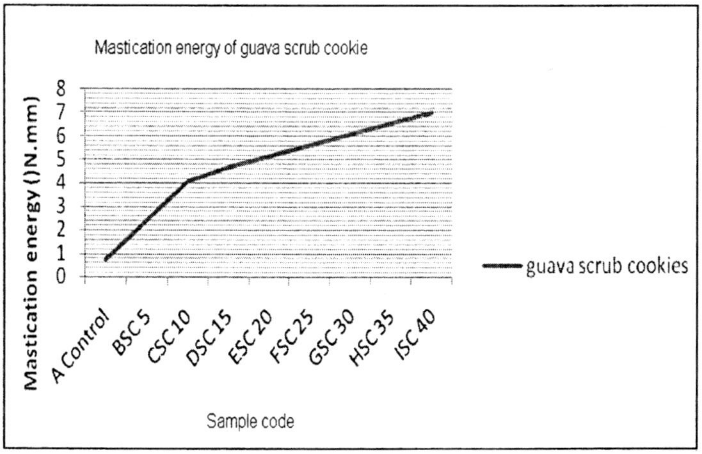

Figure 6 *Mastication energy of guava cookies*

3.6 Sensory Analysis: Hedonic Test

From all the different formulations, the cookie formulated that has the highest fiber content but with acceptable taste were subjected to sensory analysis using a hedonic test to determine the best formulation. The taste and flavor of the cookies improved on incorporation of guava solids as these cookies had typical guava flavor. With 40 % incorporation (Figure 7) of guava scrubs, the total fiber content was the highest of all cookie variation. However, there is an undesirable tooth stick residual of grainy particles after chewing.

Figure 7 *40 % guava scrub cookie variation*

Therefore, it was decided that the 30 % guava scrub incorporation, can be considered as the best variation for guava scrub cookies. As for guava seeds and leaves cookie, 40 % guava solid incorporation is the cookie formulation of the highest fiber percentage. It also has an attractive appearance, with the leaves scattered on top of the cookie before baking (Figure 8).

One way unstacked ANOVA of control cookie, GSC 30 cookie (30 % scrub) and YSE 40 LE cookie (40 % seeds and leaves) variation was carried out using the MINITAB *Release 14 for Windows* from the data obtained from the questionnaire distributed to 60 judges to determine its statistical difference of overall acceptance.

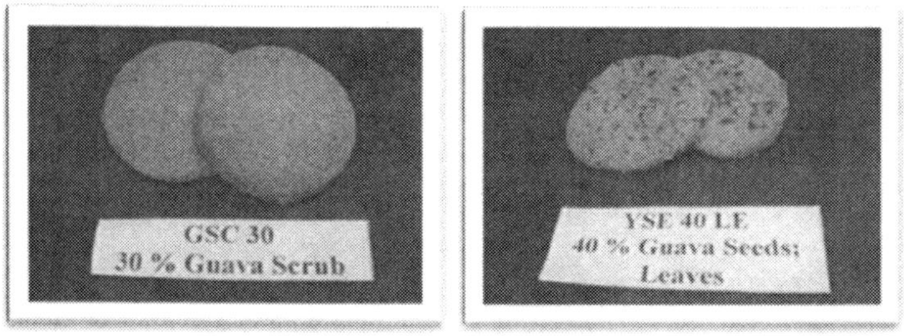

Figure 8 *30 % guava scrub cookie and 40% guava seeds and leaves cookie*

The test statistic, $F = 15.54$, has a p-value of 0.00. The p-value is the probability of obtaining a test statistic as large as $F = 15.54$, assuming H_0 is true. Since the p-value is less than $\alpha = 0.05$, the null hypothesis is rejected. We have sufficient evidence to conclude that the true mean control differ for at least one of the mean from the sample cookie variation.

To determine which means differ from each other, Tukey's Test is then carried out. The point estimate for the mean for GSC 30 (μ_2) cookie variation subtracted from the mean for control (μ_1) is -0.100; the Tukey interval is $-0.664 \leq \mu_1 - \mu_2 \leq 0.464$; which contain a zero, indicating no differences. However, the point estimate for the mean for YSE 40 LE (μ_3) cookie variation subtracted from the mean for control (μ_1) is 1.100; the Tukey interval is $0.536 \leq \mu_1 - \mu_3 \leq 1.664$; which do not contain zero, indicating a difference. Thus, it can be concluded that the mean of control differs significantly between control and YSE 40 LE cookie. A detailed One-way ANOVA: CONTROL, GSC 30, YSE 40 LE and Tukeys' Test can be viewed at Appendix C. The mean values of each cookie variation are then inserted into a pie chart (Figure 9).

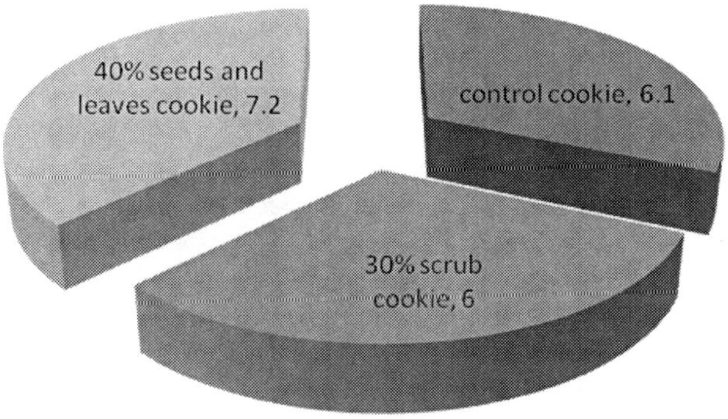

Figure 9 *Mean Value of Overall Acceptability of Cookie Formulation*

Thus, it can be concluded that the 40% guava seed and leaves cookie is the best formulation which has the highest overall acceptability, while the 30% guava scrub cookie comes in second. Considering the appearance, texture and flavor attributes, it could be inferred that 40% level of guava seeds with leaves scattered and the 30% guava scrubs was optimum.

Gutiérrez *et al.*[1] summarized that in their study, guava leaves provide antioxidant and other effects providing beneficial protective properties to the heart and liver with an improvement in myocardial and muscular function, anti-allergic, anti-inflammatory, analgesic, sedative, and central nervous system depressant activity. They also added that this is the reason why some folk recipes uses leaf extracts for rheumatic pain, convulsions, vomiting, and menstrual pain. This shows that the addition of leaves will add extra nutraceutical value to the special high fiber guava cookies formulated.

4 CONCLUSIONS

This study shows that the best variation is the 30% substitution of guava scrubs and 40% of guava seeds and leaves cookie with commercial fine flour in the cookies resulted in a highly acceptable sensory scores (mild guava taste, appealing crust colour and appearance, and crunchy mouth feel texture), and excellent storage quality. Between the two; the 40% of guava seeds and leaves cookie results in significantly different ($P \leq 0.05$) overall acceptability compared to the control, which conclude that best part of pink guava in formulating high fiber guava cookies are the seeds which are grinded and then sieved with crushed leaves scattered, and the scrubs. Therefore, pink guava solids, could be utilized for the preparation of cookies and is applicable as a bakery product ingredient and other food products with improved functional and nutraceutical properties.

References

1 S. Gorinstein, M. Zemser, R. Haruenkit, R. Chuthakora, F. Grauer, O. Martin-Belloso and S. Trakhten Berg, *J. Nutr. Biochem.,* 1999, **10**, 367.
2 A. Jiménez-Escrig, M. Rincón, R. Pulido and F. Saura-Calixto, *J. Agric. Food Chem.,* 2001, **49**, 5489.
3 P. Ramulu and P.U. Rao, *J. Food Compos. Anal.,* 2003, **16**, 677.
4 C.M. Ajila, K. Leelavathi and U.J.S. Prasada Rao, *J. Cereal Sci.,* 2008, **48**, 319.
5 W. Gisslen, *Professional baking*. 5th Ed., John Wiley & Sons Inc., New Jersey, 2009.
6 K. Leelavathi and P. Haridas Rao, *J. Food Sci. Technol.,* 1993, **30**, 187.
7 R. M. P. Gutiérrez, S. Mitchell and R.V. Solis, 2008. Psidium guajava: A review of its traditional uses, phytochemistry and pharmacology. *Journal of Ethno-Herbiotech Development Incorporation, 2006. On guava tea.*
8 M.L. Sudha, R. Vetrimani and K. Leelavathi, *J. Food Chem.,* 2007, **100**, 1365.
9 S. Hooda and S. Jood, *Food Chem.*, 2005, **90**, 427.
10 A. Rathi, A. Kawatra, S. Sehgal and B. Housewright, *LWT Food Sci. Technol.*, 2004, **37**, 187.
11 H.R. Sharma and G.S. Chauhan, *J. Food Sci. Technol.* 2002, **39**, 225.
12 M.M. Patel and V. Rao, *J. Food Sci. Technol.* 1996, **33**, 53.
13 H. Chen, G.K. Tubenthaler, H.K. Leung and J.D. Baranowski, *Cereal Chem.*, 1998, **65**, 244.
14 L. Kirssel and M. Prentice, *Cereal Chem.*, 1979, **50**, 261.
15 K.H. McWatters, *Cereal Chem.*, 1978, **55**, 853.
16 R. Singh, G. Singh and G.S. Chauhan, *J. Food Sci. Technol.* 1996, **33**, 355.

EXTRACTION OF ANTIOXIDANT COMPOUNDS FROM APPLE POMACE

Hilde Henny Wijngaard and Nigel Brunton

Teagasc Ashtown Food Research Centre, Ashtown, Dublin 15, Ireland

1 INTRODUCTION

Sustainable food production and waste valorisation have become important issues in the food industry. Apple pomace is generated globally through apple juice and apple cider production and is an example of an abundant and reasonable uniform by-product.

Apple pomace has been used as a source of numerous compounds with technological applications,[1] including pectin,[2] citric acid,[3] ethanol[4] and enzymes.[5,6] Apple pomace could also serve as a source compounds with potential nutritional and technological benefits such as antioxidative polyphenols.[7] The use of apple pomace as a source of polyphenols for inclusion in other foods would require the use of food grade solvents, such as ethanol and water, as the extracting media. These solvents are also more environmental friendly which would also benefit the sustainability of the process. The use of so called novel "green" extraction techniques which minimise solvent use while maximising extraction yields could also deliver additional economic and sustainability benefits.

Pressurised liquid extraction (PLE) is a synonym of accelerated solvent extraction (ASE) and can enhance extraction rates of target compounds from food matrices. By applying high pressures, this technique allows the user to complete extractions at temperatures higher than the boiling point of the solvent.[8] Hereby the analyte solubility and mass transfer rates are increased, which generally leads to better recoveries of the target compounds than standard solid-liquid extraction techniques. In addition, to a greater extraction efficiency rate, PLE requires only small amounts of solvents, which makes PLE a green extraction technique.[9] Recovery of target compounds from foods is generally most influenced by solvent composition and extraction conditions. This principle applies to both conventional solid liquid extraction techniques carried out at atmospheric pressures and extractions carried out at elevated pressures as is the case for PLE.

Response Surface Methodology (RSM) is a useful technique for reducing the number of experiments required to identify optimal conditions for a target variable. The technique uses a combination of mathematical and statistical techniques, in which modelling is used to optimize a response value that is influenced by various factors. RSM has been widely used to optimise food processing conditions.

The objectives of the present study were to optimise the conventional ethanol/water extraction of polyphenols from apple pomace and test the potential of PLE for same. Since the ultimate aim was to recover these compounds for use in foods ethanol and water, two environmentally and food safe solvents were used and RSM was applied in order to maximise antioxidant and polyphenol recoveries.

2 MATERIALS AND METHODS

2.1 Preparation of materials

Apple pomace was provided by Bulmers Limited (Clonmel, Ireland). The samples were vacuum packed to prevent oxidation and fermentation and stored at -20°C for at least 24 hours. After this, samples were lyophilised for a minimum of 5 days in an A6/14 freeze dryer (Frozen in Time Ltd., York, UK) vacuum packed and stored at -20°C until required for analysis. On the day of extraction, lyophilised samples were milled to a fine powder using a blender (BL440001, Kenwood Ltd., Hampshire, UK).

2.2 Conventional Extraction Procedure

Conventional extraction was carried out using the procedure described by Wijngaard and Brunton.[10] In brief, after adding ethanol/water solution lyophilised samples were homogenised, heated in water bath, cooled down, vortexed, centrifuged and filtered.

2.3 PLE

Apple pomace was placed between two layers of diatomaceous earth, in PLE cells with a cellulose filter at either end. A temperature range between 80°C and 170°C and various ethanol:water concentrations were used in order to identify optimal conditions for antioxidant and polyphenol extraction by RSM. An accelerated solvent extractor (ASE 200, Dionex, Sunnyvale, USA) was used and a standard stepwise PLE extraction protocol was applied for all extractions. Extracts were collected in 60 mL glass vials (Dionex, Sunnyvale, USA). The samples were centrifuged for 15 min at 3,000 rpm (MSE Mistral 3000i, Sanyo Gallenkamp, Leicestershire, UK). 10 mL Of the supernatant was filtered through 0.22 μm PVDF syringe filters (Pall Gelman Laboratory, Portsmouth, UK). The extracts were stored at -20°C.

2.4 DPPH assay

A modified version of the DPPH method[11] was used. Prior to analysis, serial dilutions of the ethanolic extracts of the apple pomace samples were prepared. 500 μl of diluted ethanolic extracts was added to 500 μl of the DPPH working solution (0.048 mg DPPH / ml). After vortexing, the test samples were left in the dark for 30 min at room temperature. The absorbance was then measured against ethanol at 515 nm in a spectrophotometer (UV-1700 Pharma Spec, Shimadzu, Milton Keynes, UK). The decrease in absorbance of a sample was calculated in comparison to a blank sample (500 μl methanol and 500 μl DPPH). The antioxidant activity was defined as the concentration of sample extract necessary to obtain an activity of 50% (IC50). The final results for antioxidant activity were determined using the following equation: antioxidant activity = (IC50Trolox / IC50Sample) x 105.[12] The antioxidant activity was expressed in mg Trolox equivalent (TE) per 100g dry weight sample (mg TE / 100 g DW).

2.5 Polyphenol analysis

2.5.1 FCR assay. Total phenolic content of ethanolic apple pomace extracts was assessed using a modified version of the Folin-Ciocalteu assay.[13] An aqueous gallic acid solution (200 mg / l) was used to create a standard curve. For the analysis, 100 µl of sample extract or gallic acid standard, 100 µl of methanol, 100 µl of Folin-Ciocalteu reagent and 700 µl of Na_2CO_3 were aliquoted into a 1.5 ml micro-centrifuge tube. The samples were vortexed immediately and the tubes were incubated in the dark for 20 min at room temperature. After incubation all samples were centrifuged at 13,000 rpm for 3 min. The absorbance of the supernatant was then measured at 735 nm using a spectrophotometer (UV-1700 Pharma Spec, Shimadzu, Japan). The results were expressed in mg gallic acid equivalent / 100 g dry weight (mg GAE / 100 g DW).

2.5.2 HPLC-DAD. HPLC-analysis was performed on a Varian Pro Star (Varian Inc., Walnut Creek, USA) chromatograph system, equipped with a module 335 diode array detector with an absorbance detection range between 190 to 950 nm. Separations were conducted on a Zorbax SB C18, 5µm, 150 x 4.6mm column (Agilent Technologies, Dublin, Ireland). The gradient profile was based on a method of Tsao and Yang.[14] Acetic acid in 2 mM sodium acetate (final pH 2.55, v/v) was used as eluent A and 100% acetonitrile was used as eluent B; a gradient profile was used. Hydroxybenzoic acids, dihydrochalcones, procyanidins and flavanols were monitored at a wavelength of 280 nm, hydroxycinnamic acid derivatives at 320 nm, and flavonols at 360 nm. For quantification and identification purposes standard curves of analytes of interest were prepared using authenticated standards. Levels of chlorogenic acid were expressed in µmol chlorogenic acid / 100 g DW apple pomace, total flavonol level was expressed in µmol rutin / 100 g DW apple pomace and the level of phloretin glycosides was expressed in µmol phlorizin / 100 g DW apple pomace.

2.6 Experimental Design

Response surface methodology (RSM) was used to optimise antioxidant activity and polyphenol extraction from apple pomace. A Central Composite design was used for both conventional extraction and PLE. The independent variables were coded at five levels ($-\alpha$, -1, 0, 1, α) and was using five central points. Design Expert 7.1.3 (Stat-Ease Inc., Minneapolis, USA) was used for designing experiments and statistical data analysis (ANOVA). Experimental data were analysed to fit a second order polynomial model. Models and regression coefficients were considered significant when p-values were lower than 0.05.

3 RESULTS AND DISCUSSION

3.1 Conventional extraction

The main polyphenols present in solid-liquid extracts from apple pomace are chlorogenic acid, flavonols, and phloretin glucosides.[10] Catechins and procyanidins and other cinnamic acids such as caffeic acid are present in very small amounts. Anthocyanins are absent from the extracts. Similar results have been reported by other authors, when extracting apple pomace with ethanol.[15] Analysis of these polyphenols, DPPH values and total phenols as response values by RSM has revealed that second order polynomial models for these responses were significant. Ethanol concentration had a significant effect on all the responses. This implies that the polarity and size of phenolic compounds play an important role in their respective extraction yields. Temperature significantly affects antioxidant activity but does not affect individual polyphenol groups, which seemed quite stable, even at temperatures of 80°C.[10]

Conditions for maximum antioxidant activity by conventional extraction as calculated by RSM are: 56% ethanol at 80°C for 31 min. Replicate extractions (n = 3) of the optimal points had antioxidant values of 444 mg Trolox / 100 g DW. Although extracting with ethanol results in lower antioxidant activity than methanolic extracts (636 mg Trolox / 100 g DW),[16] ethanol seems a reasonable replacement for methanol as a food grade and environmentally friendly solvent.

3.2 PLE

PLE enhances antioxidant activity of the extracts as illustrated in Figure 1. At a temperature of 170°C, the antioxidant activity can be greater by a factor 3 when compared to ethanolic solid-liquid extraction. A disadvantage was that browning compounds were formed. While these compounds may have enhanced the antioxidant activity, their presence at temperatures higher than 150°C implies the formation of toxic compounds, such as furans as well.[17] Therefore more work needs to be carried out to analyse ranges of temperatures below and above temperatures of 150°C for PLE extraction.

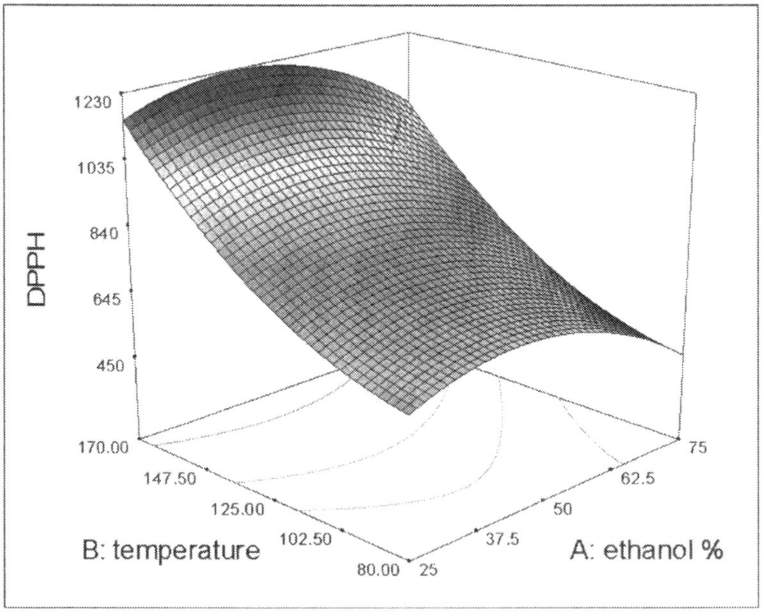

Figure 1 *Response Surface plot of PLE showing the effects of temperature and ethanol concentration on antioxidant activity as measured by DPPH (mg TE/100 g DW).*

A similar polyphenol profile for PLE extracts as compared to solid-liquid extracts was obtained i.e.,chlorogenic acid, flavonols, and phloretin glucosides were present. Using optimal extractions yields of respective phenol groups were enhanced by at least a factor of two in comparison to conventional solid-liquid extraction.

4 CONCLUSIONS

- Ethanol concentration had a significant affect on extraction yields of all antioxidant indices for both PLE and solid-liquid methods
- Antioxidant extraction yields for PLE were several fold higher than for solid-liquid extraction. However the formation of browning compounds at the higher temperatures required for these yields may limit the usefulness of these extracts because of toxicity problems.

References

1 F. Vendruscolo, P.c.M. Albuquerque, F. Streit, E. Esposito and J.L. Ninow, *Crit. Rev. Biotechnol.*, 2008, **28**, 1.
2 A. Schieber, P. Hilt, P. Streker, H.U. Endress, C. Rentschler and R. Carle, *Inn. Food Sci. Emerg. Technol.*, 2003, **4**, 99.
3 Y.D. Hang and E.E. Woodams, *Biotechnol. Lett.*, 1984, **6**, 763.
4 Y.D. Hang, C.Y. Lee, E.E. Woodams and H.J. Cooley, *Appl. Environ. Microbiol.*, 1981, **42**, 1128.
5 M. Berovic and H. Ostroversnik, *J. Biotechnol.*, 1997, **53**, 47.
6 E. Favela-Torres, S.T. Volke and G.G. Viniegra, *Food Technol. Biotechnol.*, 2006, **44**, 221.
7 S. Bhushan, K. Kalia, M. Sharma, B. Singh and P.S. Ahuja, *Crit. Rev. Biotechnol.*, 2008, **28**, 285.
8 A. Beyer and M. Biziuk, *Food Chem.*, 2008, **108**, 669.
9 J.A. Mendiola, M. Herrero, A. Cifuentes and E. Ibanez, *J. Chromatogr. A*, 2007, **1152**, 234.
10 H.H. Wijngaard and N. Brunton, *J. Food Eng.*, **In Press, Accepted Manuscript**.
11 P. Goupy, M. Hugues, P. Boivin and M.J. Amiot, *J. Sci. Food Agric.*, 1999, **79**, 1625.
12 S.F. Hagen, G.I.A. Borge, G.B. Bengtsson, W. Bilger, A. Berge, K. Haffner and K. A. Solhaug, *Postharvest Biol. Technol.*, 2007, **45**, 1.
13 V.L. Singleton, R. Orthofer, R.M. Lamuela-Raventós and P. Lester in *Methods in Enzymology*, ed. L. Packer, Academic Press, 1999, Volume 299, p 152.
14 R. Tsao and R. Yang, *J. Chromatogr. A*, 2003, **1018**, 29.
15 X. Cao, C. Wang, H. Pei and B. Sun, *J. Chromatogr. A*, 2009, **1216**, 4268.
16 H.H. Wijngaard, C. Rößle and N. Brunton, *Food Chem.*, 2009, **116**, 202.
17 K.A. Tehrani, M. Kersiene, A. Adams, R. Venskutonis and N. De Kimpe, *J. Agric. Food Chem.*, 2002, **50**, 4062.

EXTRACTING NOVEL FOAM AND EMULSION STABILITY ENHANCERS FROM BREWERS' GRAIN

F.A. Husband, A. Jay, C.B. Faulds, K.W. Waldron and P.J. Wilde *

Institute of Food Research, Norwich Research Park, Colney, Norwich, NR4 7UA. UK

1 INTRODUCTION

Brewers' grain (BG) is a co-product of the brewing industry and is often disposed of as animal feed, compost and into landfill. Reduced animal numbers and increased landfill tax mean rising disposal costs, thus creating opportunities for exploiting BG.

One important quality attribute of beer is the appearance of a good, stable foam.[1] Propylene glycol alginate (PGA) is sometimes used to enhance beer foam stability, but may have to be declared on packaging following recent and proposed EU food labelling legislation. Therefore a more natural alternative to PGA is being sought. Beer foam is stabilised by protein, which forms a viscoelastic "skin" on the surface of foam bubbles.[2,3] The strength of the protein film can be broken down by lipids and surfactants,[4] thus destabilising the foam.[5] PGA improves foam stability possibly via interactions with the protein film,[6,7] increasing the stability of the interfacial layer, and perhaps also slowing the drainage of liquid in the foam lamellae due to local viscosity increases. Arabinoxylans (AX) displayed similar foam protection properties to PGA,[8] but the molecular mechanisms were not entirely clear.

BG contains extractable AX -rich fractions.[9] Degrees of substitution of AX differ between wheat and barley, and the presence and nature of the substitutions are also are affected by the malting process.[9] Fleury[10] demonstrated that environmental and genetic factors have also been shown to affect extractable arabinoxylan. Therefore BG is potentially a source of AX fractions that could replace PGA as a foam enhancing agent.

A similar approach could also be applied to emulsions, to extend physical stability and prolong shelf life. Many existing products used to enhance emulsion stability such as gelatine, carageenan, gum arabic, have various drawbacks regarding product consistency and reliability of supply. Also increasingly important to consumers is the carbon-footprint of foods from gate to plate.

Arabinoxylan extracted from spent grain could meet these functional and environmental requirements, therefore the aim of this study was to investigate the efficacy and mechanism of action of AX fractions extracted from BG to provide an alternative and more natural source of foam and emulsion stabilising ingredients. The initial focus was on beer foam, and the potential of AX to replace PGA. We also present preliminary results on emulsion stability to demonstrate the versatility of the AX extracts.

2 METHODS AND RESULTS

2.1 Materials

Brewers spent grain was kindly supplied by a local brewer. De-ionised water was used throughout. Tween 20 (product no. 28320) was purchased from Pierce (Cramlington, UK). All other chemicals were supplied by Sigma (Poole, UK).

2.2 Methods

2.2.1 Production of arabinoxylan –rich fractions. AX rich fractions were extracted from BG as reported previously.[9] In brief, the BG was de-watered using a commercial spin dryer, then freeze dried, and roughly ground using a pestle and mortar. This material was sieved through a 1mm sieve, and ground to fine powder using a laboratory mill. 20g of ground fraction was then alcohol extracted as follows: powder was stirred in 1L 70%(v/v) ethanol at 20°C for 5 minute, the sample was sieved through a nylon 30μm mesh. This procedure was then repeated with firstly 70%(v/v) ethanol at 80°C for 5 minutes, then 99.5%(v/v) ethanol at 78°C for 5 minutes. The solid filter cake was then washed with 250ml ethanol, followed by 250ml acetone. The powder was allowed to air-dry in the fume hood. A sequential extraction of the powder was then followed as per Mandalari *et al,* [9] and summarised in Figure 1.

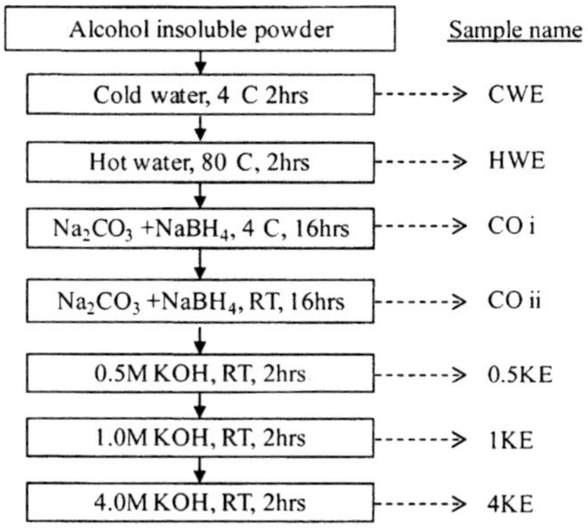

Figure 1 *Summary of the BG sequential extraction process*

2.2.2 Foam Stability. The foam stability was determined by two methods. A micro-conductimetric method as previously described [11] was used to assess the model systems. Here, 1.8ml of solution was sparged through a single 30μm orifice. The conductivity of the foam was monitored using flat platinum electrodes. The foam stability was calculated by expressing the foam conductivity after 5 minutes drainage as a percentage of the initial foam conductivity. A model system designed to mimic the behaviour of a lipid destabilised beer foam was developed and comprised of - 4mg/ml β-lactoglobulin (BLG), 12.5μM Tween 20 in 20mM citric acid/40mM disodium hydrogen phosphate, at pH4.2, with 4% ethanol. AX fractions and PGA were added to quantify their foam enhancing capacity. A similar approach was also used on a commercial, PGA-free beer sample, using the industry standard RUDIN test.[12] 45ml of degassed beer was foamed by sparging through a sintered frit to a fixed height (325mm). The time taken for the drained liquid to rise from 50 to 75mm up the tube was measured and taken as a half-life of the foam drainage (Head Retention Value – HRV).

2.2.3 Surface properties. Surface and interfacial tension was measured using the pendant drop technique (FTÅ200 - First Ten Angstroms, Virginia). The size and shape of a drop suspended from a teflon coated needle was used to calculate the surface tension from the Laplace-Young equation.

The surface dilational elastic modulus (E) was calculated by oscillating the pendant drop volume sinusoidally and the resulting changes in surface area and surface tension were used to calculate the surface rheological parameters, as described previously.[13]

Interfacial shear rheology was carried out using an AR2000 controlled stress rheometer (TA Instruments, Crawley, UK),[14] using an aluminium bicone (6°). The measurements were performed under low deformation, oscillation at an amplitude of 1% and frequency of 0.5Hz to minimise any impact on the integrity of the interfacial layer.

2.2.4 Emulsion stability. 3ml of a 10 %v/v oil in water emulsion, stabilised by 1.5mg/ml β-casein were emulsified using a bench top homogeniser (Emulsiflex B-3, Avestin Inc. Ottowa, Canada). The emulsions circulated around a small volume chamber of a laser diffraction sizer (LS-230, Coulter) to induce coalescence. The droplet size distribution was monitored every minute to quantify the extent of coalescence in the presence and absence the 1KE AX extract.

2.3 Results

2.3.1 Foam stability.

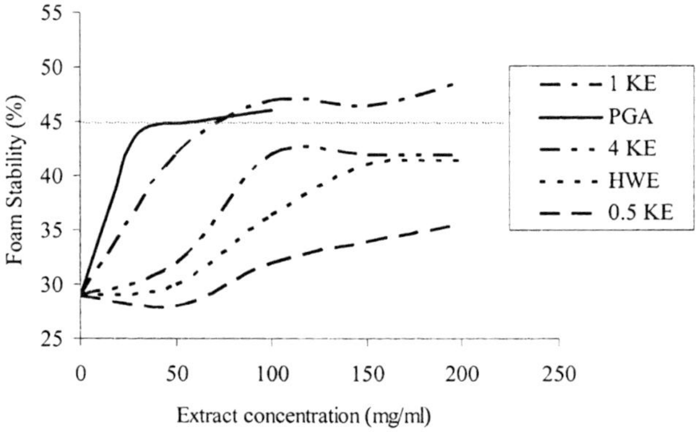

Figure 2 *Foam stability of BLG/Tween 20 system as a function of extract type and amount. HWE(• • • •); 0.5 KE (– – –); 1 KE (– • –); 4 KE (– • •); PGA (—).*

Foam restabilisation of a destabilised model systems by the arabinoxylan-rich (AX+) fractions are shown in Figure 2. Without the addition of Tween 20 the foam stability of BLG was 45%, as shown by the light, dotted line. Addition of Tween 20 destabilised the foam down to approximately 30%, as reported elsewhere.[15] The addition of the most effective AX fractions increased the foam stability. The 1KE fraction was the most effective and was roughly comparable to that observed for PGA.

Further tests were performed using the industry standard RUDIN test on commercial beer that was also free from foam enhancers such as PGA. In the absence of Tween 20, both PGA and 1KE showed similar improvements in HRV as shown in Figure 3. In the presence of Tween 20 the recovery in foam stability was not as great as the microconductivity results (Fig. 2). However both the PGA and 1KE produced similar levels of enhancement in the HRV values supporting the possibility that the 1KE fraction could potentially be a viable PGA replacement

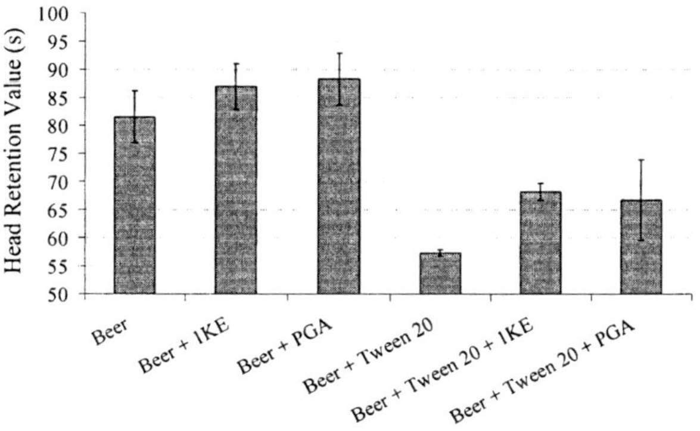

Figure 3 *HRV values from Rudin test of 100 µg/ml 1KE and 50µg/ml PGA added to beer in the presence and absence of 12.5µM Tween 20.*

2.3.2 Surface properties. The surface shear rheological moduli (Elastic - G' and viscous - G'') of the model system was determined.

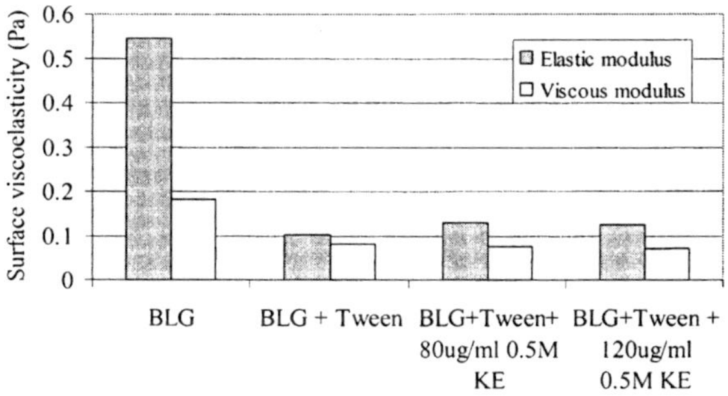

Figure 4 *Surface elastic and viscous moduli of 0.3 mg/ml BLG, with added Tween 20 (0.5µm) and with added 0.5 KE fraction.*

The hypothesis was that the enhancement of foam stability was due to the AX interacting with adsorbed protein, strengthening the surface, which is likely to reduce the impact of the adsorbing lipids or surfactants.[16] The presence of Tween 20 causes a reduction in both G' and G'', as would be expected. However, the addition of 1KE to BLG / Tween 20 caused no further increase in either G' or G''. The results are shown in Figure 4. A range of arabinoxylan concentrations were tested, all showing no further increase in G' or G'' (data not shown).

2.3.3 Emulsion stability. Emulsion stability was determined by monitoring the extent of coalescence over a period of 9 minutes. Dephosphorylated β-casein stabilised emulsions are sensitive to shear,[17] and coalescence occurs readily as shown in figure 5. However, in the presence of 1KE, coalescence appears to have almost completely stopped over the time scale of the experiment. Interfacial tension and elasticity measurements were similar to those shown in Figure 4, showing that the AX fraction did not strengthen the adsorbed protein layer.

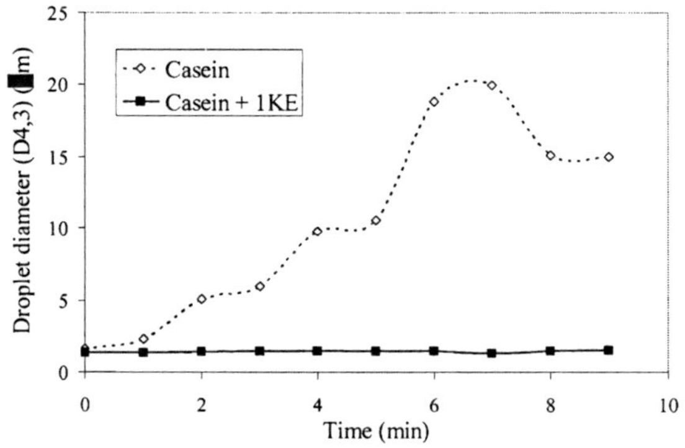

Figure 5 *Mean droplet diameter ($D_{4,3}$) of n-tetradecane emulsions stabilised by dephosphorylated β-casein in the presence and absence of 2.5 mg/ml 1 KE.*

3 CONCLUSIONS

Arabinoxylan (AX) rich fractions extracted from the co-product - Brewers' Grain are able to enhance the stability of protein stabilised foams and emulsions. They are able to improve beer foam stability to a similar efficacy of an existing foam enhancer (PGA). The most active fraction was that extracted using 1M KOH. Our working hypothesis was that the AX fractions were able to strengthen the protein adsorbed layer, making them more resistant to the damaging effects of lipids and surfactants. However, surface and interfacial measurements suggested that no such strengthening occurs. Further work on the interfacial properties of the adsorbed films in the presence of the AX fraction is required, to study the effects on thickness and charge.

Acknowledgement

This work was supported by Biotechnology and Biological Sciences Research Council (BBSRC), UK, and Department of the Environment, Food and Rural Affairs (DEFRA), through a BridgeLink Grant (AFM 201) and LINK (AFM). We would like to express our gratitude to Scottish Courage Limited, Edinburgh, UK for supplying brewers' spent grain.

References

1 C.W. Bamforth, *J. Inst. Brew.*, 2000, **106**, 229.
2 D.C. Clark, R. Dann, A.R. Mackie, J. Mingins, A.C. Pinder, P.W. Purdy, E.J. Russell, L.J. Smith and D.R. Wilson, *J. Colloid Interface Sci.*, 1990, **138**, 195.
3 E. Dickinson, 'Protein adsorption at liquid interfaces and the relationship to foam stability' in *Foam: Physics, Chemistry and Structure,* ed., A.J. Wilson, Springer-Verlag, Berlin, 1989, Chapter 4, pp. 39-53.
4 J-L. Courthaudon, E. Dickinson, Y. Matsumura and D.C. Clark, *Colloid Surf.*, 1991, **56**, 293.
5 D.J. Cooper, F.A. Husband, E.N.C. Mills and P.J. Wilde, *J. Agric. Food Chem.*, 2002, **50**, 7645.
6 G. Jackson, R.T. Roberts and T. Wainwright, *J. Inst. Brew.*, 1980, **86**, 34.
7 D.K. Sarker and P.J. Wilde, *Colloid Surf., B*, 1999, **15**, 203.
8 D.K. Sarker, P.J. Wilde and D.C. Clark, *Cereal Chem.* 1998, **75**, 493.
9 G. Mandalari, C.B. Faulds, A.I. Sancho, A. Saija, G. Bisignano, R. LoCurto and K.W. Waldron, *J. Cereal Sci.*, 2005, **42**, 205.
10 M.D. Fleury, M.J. Ednay, L.D. Campbell and G.H. Crow, *Can. J. Plant Sci.*, 1997, **77**, 191.
11 D.C. Clark, P.J. Wilde and D.R. Wilson, *Colloid. Surf.*, 1991, **59**, 209.
12 P.K. Hegarty, 'Beer Foam' in *Foam: Physics, Chemistry and Structure,* ed., A.J. Wilson, Springer-Verlag, Berlin, 1989, Chapter 12, pp. 197-206.
13 J. Maldonado-Valderrama, V.B. Fainerman, M.J. Gálvez-Ruiz, A. Martín-Rodriguez, M.A. Cabrerizo-Vílchez and R. Miller, *J. Phys. Chem. B*, 2005, **109**, 17608.
14 A.R. Mackie, M.J. Ridout, G. Moates, F.A. Husband and P.J. Wilde, *J. Agric. Food Chem.*, 2007, **55**, 5611.
15 M. Coke, P.J. Wilde, E.J. Russell and D.C. Clark, *J. Colloid Interface Sci.*, 1990, **138**, 489.
16 A.R. Mackie, A.P. Gunning, P.J. Wilde, and V.J. Morris, *J. Colloid Interface Sci.*, 1999, **210**, 157.
17 F.A. Husband, P.J. Wilde, A.R. Mackie and M.J. Garrood, *J. Colloid Interface Sci.*, 1997, **195**, 77.

BIOLOGICAL PRODUCTION OF VANILLIN FROM FERULIC ACID OBTAINED FROM WHEAT BRAN HYDROLYZATES

D. Di Gioia,[1,2] L. Sciubba,[1] M. Ruzzi[3] and F. Fava[1]

[1] DICASM, Faculty of Engineering, *Alma Mater Studiorum* - University of Bologna, Bologna, Italy
[2] Department of Agroenvironmental Science and Technologies, *Alma Mater Studiorum* - University of Bologna, Bologna, Italy
[3] Department of Agrobiology and Agrochemistry, University of La Tuscia, Viterbo, Italy.

1 INTRODUCTION

The valorization of by-products of agricultural origin is part of a sustainable development approach. Wheat bran, an abundant agro-industrial by-product in Italy, contains about 5 g/kg of dry matter of ferulic acid,[1] which can be released from cell wall polysaccharides through chemical or enzymatic hydrolysis.[2,3] Ferulic acid is a precursor of vanillin, one of the most used flavouring compound in the food industry.[4] There are several examples in the literature of biotechnological production of vanillin starting from this substrate,[5–8] but only a few studies are focused on the obtainment of vanillin from industrial by-products such as cereal bran.[2,9] A recent work has shown that ferulic acid occurring in wheat bran hydrolyzates can be successfully bioconverted into vanillin by resting cells of *Escherichia coli* JM109/pBB1 and that the purification of ferulic acid from the hydrolyzates can increase vanillin molar yield.[2] In this work the role of ferulic acid concentration in wheat bran hydrolyzates used as ferulic acid source is explored in order to evidence possible inhibitory effect on the bioconversion process by high substrate concentration. Moreover, different ferulic acid purification methods are compared in terms of vanillin molar yields obtained by the bioconversion of the recovered ferulic acid. In addition, the possibility of exploiting the carbohydrate rich water phase obtained after ferulic acid recovery as the growth substrate for the microorganism operating the bioconversion is explored.

2 METHOD AND RESULTS

2.1 Bioconversion of crude wheat bran hydrolyzates amended with food grade ferulic acid

E.coli JM109/pBB1, containing a plasmid derivative with a catabolic cassette for the conversion of ferulic acid into vanillin, was used as the microbial agent for the bioconversion.[5] Wheat bran hydrolyzate preparation, ferulic acid hydrolysis with the use of a mixture of commercial enzymes, and the procedure for bioconversion experiments by *E.coli* JM109/pBB1 resting cells were previously described.[2] The concentration of vanillin, ferulic acid and other bioconversion products was evaluated through reverse phase HPLC-DAD analysis.[2]

Crude wheat bran hydrolyzates had a ferulic acid concentration of about 230 mg/L and contained about 70g/L of total sugars and 30g/L of reducing sugars. They were employed in the same conditions used in the previous work (i.e. undiluted, diluted 3:1 v/v and 1:1 v/v in distilled water),[2] but each sample was amended with 0.5 mM (100 mg/L) food grade ferulic acid, thus obtaining initial concentrations of ferulic acid of 1.5 mM, 1.25 mM and 1 mM, respectively. The aim of this experiment was to study the effect of higher substrate concentration with respect to those present in not-amended hydrolyzates on the bioconversion process, the sugar concentration being constant. The vanillin molar yield obtained in this experiment are shown in Figure 1 and are very similar to those obtained in the absence of exogenous ferulic acid.[2] In particular, vanillin produced from ferulic acid was quickly reduced to vanillyl alcohol therefore contributing to the low bioconversion yield. The maximum vanillin concentration obtained in this set of experiments was 0.59 mM starting from 1.25 mM of ferulic acid after 4 h of bioconversion (not shown), whereas the maximum vanillin molar yields of these experiments were about 27%, 51%, 53% starting from 1.50, 1.25 and 1.00 mM of ferulic acid respectively. The results obtained suggest that the high concentration of ferulic acid may not be the cause of the lower yield and confirm the conclusion previously drawn on the negative role of carbohydrates present in the hydrolyzate on the bioconversion process. Therefore, successive experiments were aimed at the purification of the hydrolyzates, in order to recover ferulic acid and remove the reducing carbohydrates.

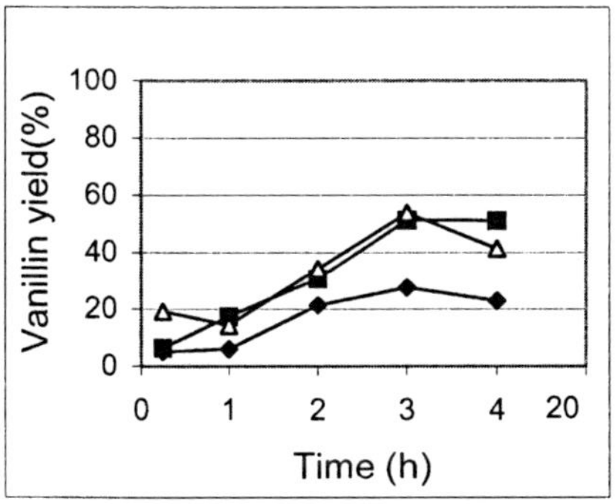

Figure 1 *Vanillin molar yields obtained using wheat bran hydrolyzates amended with 0.5 mM (100 mg/L) food grade ferulic acid, starting from 1.00 mM (△), 1.25 mM (■) and 1.50 mM ferulic acid (◆).*

2.2 Bioconversion of purified ferulic acid

Three different methods were employed to recover ferulic acid from wheat bran hydrolyzates: i) solid phase extraction with ion exchange resins Amberlite IRA® 95, as described in Di Gioia *et al.*;[2] ii) liquid-liquid extraction with ethyl acetate, herewith detailed: crude hydrolyzate was extracted twice with an equal volume of ethyl acetate, shaken and centrifuged for 5 min. at 3000 rpm. The organic phase was evaporated and the recovered ferulic acid was re-suspended in saline buffer for using in bioconversion experiments; iii) solid-liquid extraction with ISOLUTE ENV+® cartridges (Biotage, Uppsala, Sweden), herewith detailed: the columns (20 mL volume, 1 g of packing) were first washed with 9 mL of methanol and 9 mL of distilled water, then the crude hydrolyzate was applied. The water phase eluted from the column was discarded. Elution of ferulic acid was obtained with 12 mL of absolute ethanol. After solvent evaporation, ferulic acid was re-suspended in saline buffer in order to obtain the desired concentration. The recovery of ferulic acid with the three different techniques employed was 80%, 75% and 95% by using the Amberlite IRA® 95 resin, ethyl acetate and the ISOLUTE ENV+® cartridges, respectively. Therefore, these results evidenced that the most efficient recovery method is solid phase extraction with the hydrophobic cartridges.

Purified ferulic acid was used at the initial concentrations of 1.00, 0.50 and 0.25 mM for the bioconversion experiments. The data regarding the bioconversion of ferulic acid purified with Amberlite® IRA 95 are not presented, as they are quite similar to those already described in Di Gioia *et al.*[2] The best results of this set of experiments were obtained by the use of ferulic acid recovered via liquid-liquid extraction with ethyl acetate (Figure 2); the maximum vanillin molar yields were about 90%, 80% and 70% starting respectively from 0.25, 0.50 and 1.00 mM of purified ferulic acid, respectively.

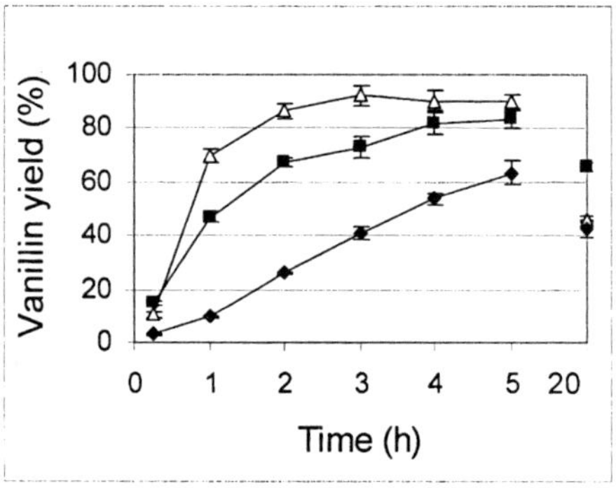

Figure 2 *Vanillin molar yields obtained using ferulic acid recovered with solvent extraction with ethyl acetate, starting from 0.25 mM (△), 0.50 mM (■) and 1.00 mM ferulic acid (◆).*

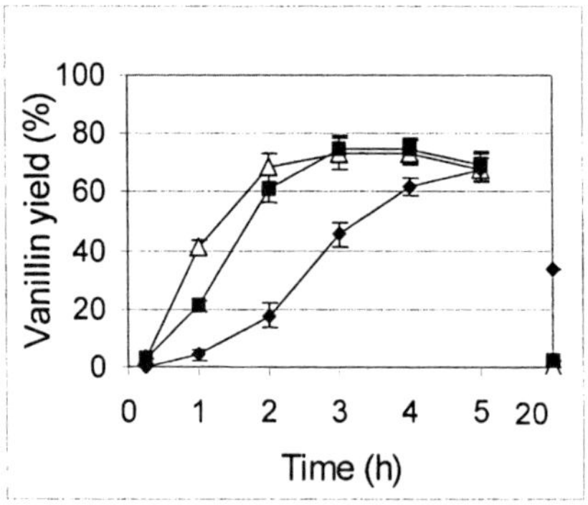

Figure 3 *Vanillin molar yields obtained using ferulic acid recovered with* ISOLUTE ENV+®, *starting from 0.25 mM (△), 0.50 mM (■) and 1.00 mM ferulic acid (◆).*

The maximum molar yields obtained with ferulic acid recovered through solid phase extraction with ISOLUTE ENV$^{+®}$ (Figure 3) were about 75%, 74% and 67% starting from 0.25, 0.50 and 1.00 mM of purified ferulic acid, respectively. The yields obtained in these experiments are higher with respect to those obtained with the crude hydrolyzate and with ferulic acid recovered by using Amberlite resins.

2.3 Use of the exhausted aqueous phase as growth substrate for *E. coli* JM109/pBB1

The aqueous phase eluted from the ISOLUTE ENV$^{+®}$ cartridges after the application of crude hydrolyzate and containing about 70 g/L of total sugars (30 g/L of reducing sugars) was sterilized through filtration and assayed as growth medium for *E. coli* JM109/pBB1. The water phase was employed as follows: (A) water phase as it was obtained after ferulic acid extraction, i.e. without any pH correction of (which was about 6), or nutrient amendments, (B) water phase with pH corrected to 7.0 with 2N NaOH prior to sterilization, (C) water phase with pH corrected to 7.0 and amended with 0.5 g/L of yeast extract. LB-grown cells were used as a positive control. Results shown in Figure 4 clearly evidenced that *E.coli* JM109/pBB1 was able to grow on the three different matrixes; the best results were achieved with the water phase amended with 0.5 g/L of yeast extract and pH corrected to 7.0, which allowed to reach an A_{600} of about 2.1 after 8 hours of growth.

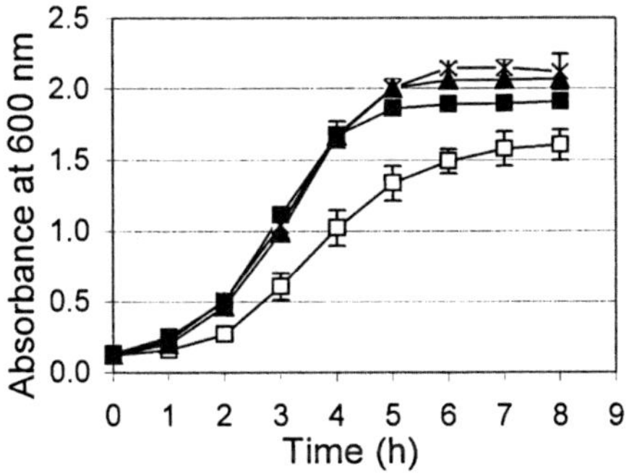

Figure 4 *Growth curve of E. coli JM109/pBB1 on the three matrixes derived from the exhausted water phase and on LB, i.e. A) not amended water phase (□); B) not amended water phase with pH corrected to 7.0, (■); C) water phase amended with 0.5 g/L of yeast extract with pH corrected to 7.0 (▲); LB, (X)*

2.4 Bioconversion experiments by using *E. coli* JM109/pBB1 cells grown on the aqueous phase derived from ferulic acid extraction

Considering the good growth capabilities of *E. coli* JM109/pBB1 on the water phase resulting from solid-liquid extraction with ISOLUTE ENV$^{+®}$ cartridges, resting cells of this strain grown on the three different matrixes were used for bioconversion experiments which employed ferulic acid recovered with the same cartridges as the substrate. Cells were grown up to a mid-late exponential phase, corresponding to an A_{600} of 1 for the first condition, an A_{600} of 1.5 for the second one and an A_{600} of 1.4 for the third one; then they were used for the bioconversion as described above. Ferulic acid was used at the initial concentration of 0,5 mM. The vanillin molar yield obtained with the three different matrixes derived from the exhausted water phase are depicted in Figure 5. Ferulic acid was bioconverted into vanillin in all the experiments. The maximum vanillin molar yield were obtained at the 7th h of incubation and were 50%, 44% and 46% for cells grown in the not-amended water phase, cells grown on the pH corrected matrix and cells grown in the matrix amended with yeast extract, respectively.

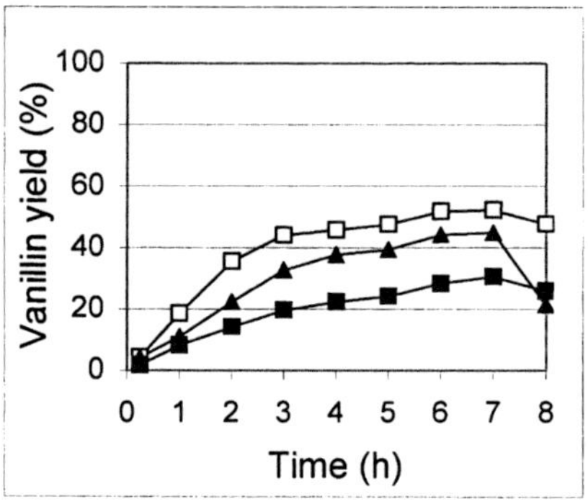

Figure 5 *Vanillin molar yields obtained from the bioconversion of ferulic acid (initial concentration 0.5 mM) recovered with ISOLUTE ENV$^{+®}$ by E. coli JM109/pBB1 grown on the three water phase derived matrixes: i.e. A) not amended water phase (□); B) not amended water phase with pH corrected to 7.0, (■); C) water phase amended with 0.5 g/L of yeast extract with pH corrected to 7.0 (▲)*

3 CONCLUSIONS

The data obtained have clearly demonstrated that the high carbohydrate concentration present in crude wheat bran hydrolyzates contribute to the low vanillin yields rather than the high concentration of ferulic acid. Therefore, purification of ferulic acid from carbohydrates appeared to be the right approach to follow.

Purification of ferulic acid with different recovery methods allowed in fact to increase the yields of the bioconversion process. The best ferulic acid recovery method was found to be solid-liquid extraction with ISOLUTE ENV$^{+®}$ cartridges as it allowed the maximum ferulic acid recovery (95%) and good bioconversion yields (up to 75%); moreover the water phase rich in reducing carbohydrates resulting from this extraction is a good growth substrate for *E.coli* JM109/pBB1. Finally, resting cells of *E.coli* JM109/pBB1 grown on this water phase are able to bioconvert the ferulic acid purified with ISOLUTE ENV$^{+®}$ into vanillin with yields of about 50% starting from 0.5 mM of substrate.

The approach described in this work allowed to obtain the complete valorisation of wheat bran components (ferulic acid plus carbohydrates) with the production of an added value natural ingredient for the modern food industry.

References

1 C.B. Faulds, G. Mandalari, R.B. Lo Curto, G. Bisignano and K.W. Waldron, *Appl. Microbiol. Biotechnol.*, 2004, **64**, 644.

2 D. Di Gioia, L. Sciubba, L. Setti, F. Luziatelli, M. Ruzzi, D. Zanichelli and F. Fava, *Enzyme Microb. Technol.* , 2007, **41**, 498.
3 H. Barberousse, A. Kamoun, M. Chaabouni, J-M Giet, O. Roiseaux, M. Paquot, C. Deroannea and C. Blecker, *J. Sci. Food. Agric.*, 2009, **89**, 1634.
4 P. Xu, D. Hua and C. Ma, *Trends Biotechnol.,* 2007, **25**, 571.
5 H. Priefert, J. Rabenhorst and A. Steinbüchel *Appl. Microbiol. Biotechnol.*, 2001, **56**, 296.
6 P. Barghini, D. Di Gioia, F. Fava and M. Ruzzi, *Microb. Cell Fact.* , 2007, **6,** 13.
7 N.J. Walton, M.J. Mayer and A. Narbad, *Phytochem.*, 2003, **65**, 505.
8 C.H. Ko, P.N. Chiang, P.C. Chiu, C.C. Liu, C.L. Yang and I.L. Shiau, *J. Chem. Technol. Biotechnol.*, 2002, **77**, 1323.
9 H.D. Shin, S. McClendon, T. Le, F. Taylor, R.R. Chen *Biotechnol. Bioeng.*, 2006, **95**, 1108.

METHANOLIC EXTRACT OF *CISTUS LADANIFERUS* AS A SOURCE OF PHENOLIC ANTIOXIDANTS FOR USE IN FOODS

M. Amensour,[1] M. Viuda-Martos,[2] E. Sendra,[2] J. Abrini,[1] J.A. Pérez-Álvarez[2] and J. Fernández-López*[2]

[1]Laboratory of Biology and Health, Department of Biology, Faculty of Science, Abdelmalek Essaadi University, Tetouan, Morocco.
[2]Grupo Industrialización de Productos de Origen Animal (IPOA). Grupo 1 UMH. Grupo REVIV. Departamento de Tecnología Agroalimentaria. Universidad Miguel Hernández. Orihuela. Alicante.

1 INTRODUCTION

Nowadays, the intensive research on healthy habits for world population has promoted a progressive increase on the production of natural products, as plants extracts as sources of bioactive substances. Many plants have been identified as having potential antioxidant activities and their consumption has been recommended.[1–4] *Cistus ladaniferus*, is a member of the family Cistaceae named rockrose which is widespread in Portugal, Spain, Italy and the North of Africa (Algeria, Morocco). The oleoresin of *C. ladaniferus* (labdanum) is appreciated by perfumists and flavourists for its tenacity and scent of amber.[5] The biological activities of Cistus extracts are widely reported and they have shown to possess antibacterial, antifungal, anti-inflammatory,[6–9] anti-aggregant of platelets,[10] anti-diarrhoeal, antiacid and antispasmodic effects.[11] This specie is a very aromatic plant because of the high essential oil content in its leaf, flower, and fruit glands.[12] The essential oil of *Cistus ladaniferus* is extremely complex. In some cases, up to 300 compounds have been detected by GC, most of them being only present as traces.[13] The GC chromatogram of the oil from *C. ladaniferus* van maculatus from Morocco showed more than 50 components among which 30 were identified. The main components were the monoterpenoids (bornyl acetate (5.5%) and pinocarveol (7.7%), and sesquiterpenoid alcohols (viridiflorol (7.2%) and ledol (3.5%)) (Oller-López, Juan et al). Furthermore, this plant is rich in flavonoids[14–16] and phenolic compounds.[17]

2 METHOD AND RESULTS

2.1 Preparation of extracts

Leaves from *C. ladaniferus* were extracted by maceration at 25% (w/v) with methanol at room temperature for 7 days. The extract was filtered and the solvent was eliminated using a rotary evaporator (Buchi Heating Bath B-490, Buchi Rotvapor R-200) to obtain a dry extract. The extract was stored at -20 °C until use.

2.2 Determination of total phenolic compounds and total flavonoids content

Total soluble phenolics in methanol extract of *Cistus ladaniferus* were determined by Folin–Ciocalteu method.[18] Quantitative measurements were performed, based on a standard calibration curve of gallic acid in methanol. The mean (±SD) results of triplicate analyses were expressed as gallic acid equivalents (GAE) in mg/gram of dry extract. Total flavonoid content was determined by using a method described by Sakanaka, Tachibana, and Okada (2005).[19] The mean (±SD) results of triplicate analyses were expressed as rutin equivalents (RE) in mg/gram of dry extract. The amount of total phenolic compounds and total flavonoids in the methanol extract of leaves of *C. ladaniferus* are shown in Table 1. Results showed that the extract is a good source of phenolic (18.43 mg gallic acid equivalents/g of dry sample) and flavonoid compounds (64.33 mg rutin equivalents/g of dry extract).

Table 1 *The content of phenolic and the content of flavonoids in methanol extract of Cistus ladaniferus leaves*

Methanol extract	Total phenolic (mgGAE/g of extract)	Total flavonoids (mgRE/g of extract)
	18.43±2.74	64.33±0.44

2. 3 Chemical composition

Analysis of the extract was performed using a Shimadzu GC-17A equipped with a Shimadzu GCMSQP5050A mass selective detector and a HP-5 MS capillary column. The components were identified by comparing their relative retention times and mass spectra with those of standards (for the main components): Wiley 229 library data of the GC/MS system, Kovats Index and literature data. All the tests were performed in triplicate.
Table 2 shows the main constituents found in the methanol extract from leaves of *Cistus ladaniferus*. In the methanol extract 15 components were identified. The major compounds were viridiflorol (38.38%), blumenol (23.17%), limonene (14.70%) and spathulenol (8.37%).

2.4 Inhibition of lipid peroxidation of buffered egg

The method of Kuppusamy, Indran, and Balraj (2002)[20] was modified, to determine the ability of the extract to inhibit peroxidation of phospholipids present in egg yolk and thus to assess the potential of the extract as a source of natural antioxidants. The inhibitory effect of methanol extract from *C. ladaniferus* on thiobarbituric acid reactive substance (TBARS) production in egg yolk induced by Fe^{2+} as shown in Figure 1.
When different concentrations of the extract (100 µg/ml – 5000 µg/ml) were added, we observed a significant concentration-dependent inhibition of lipid peroxidation. The concentration of methanolic extract that caused 50% inhibition of lipid peroxidation of buffered egg yolk was 500 µg/ml. For all concentrations assayed, the positive control (BHT) showed higher ($P<0.05$) inhibition percentages than the extracts, ranging between 83.59% at 100 µg/ml and 93.05% at 5000 µg/ml.

Table 2 *Constituents of methanol extract and their relative percentages of total chromatogram area and Kovats Index (KI)*

Pics	Compounds	KI	% Composition
1	Alpha thujene	923	0.73
2	Sabinene	970	0.80
3	α- terpinen	1016	0.69
4	**Limonene**	**1033**	**14.70**
5	Trans beta ocimene	1058	2.07
6	Linalool	1095	1.05
7	3-terpinenol	1130	-
8	β-terpineol	1161	-
9	Borneol	1171	0.71
10	Terpinen-4-ol	1180	2.50
11	Carveol	1197	0.70
12	Bornyl acetate	1285	0.69
13	Caryophyllene oxide	1587	0.75
14	**Viridiflorol**	**1599**	**38.38**
15	**Spathulenol**	**1609**	**8.37**
16	Tetradecyl aldehyde	1614	4.69
17	**Blumenol**	**1710**	**23.17**

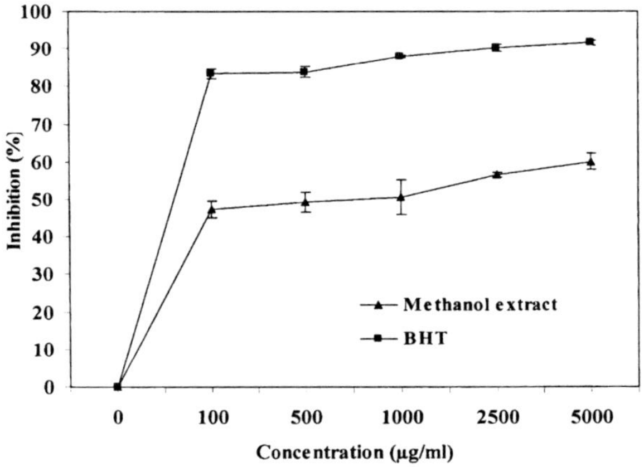

Figure 1 *Ability to inhibit lipid peroxidation of Cistus ladaniferus methanol extract compared to a control BHT*

3 CONCLUSION

The results reported are useful information that may allow the introduction of *Cistus ladaniferus* methanol extracts in food industries and the other industry processing natural products. Methanol extracts of *C. ladaniferus* exhibit a significant antioxidant activity in inhibition of TBARS test, high content of total phenolic and total flavonoids and high content on phytochemicals with proven antioxidant activities as viridiflorol, blumenol, limonene and spathulanol. However, further investigation is needed to evaluate the practical effectiveness of the extract in particular foods.

Acknowledgments - The authors would like to thank the Spanish Agency for International Cooperation (AECI) for financing this research project (A/9934/07-A/16296/08).

References

1 D.D. Kitts, Y.V. Yuan, A.N. Wijewickreme and C. Hu, *Mol. Cell. Biochem.*, 2000, **203**, 10.
2 S.E. Lee, J.H. Hyun, J. -S. Ha, H.S. Jeong and J.H. Kim, *Life Sci.*, 2003, **73**, 179.
3 F. Liu and T.B. Ng, *Life Sci.*, 2000, **66**, 735.
4 A.K. Tiwari, *Curr. Sci.*, 2001, **8**, 1187.
5 J. P. Mariotti, F. Tomi, J. Casanova, J. Costaand and A. F. Bernardini, *Flav. Fragr. J.*, 1997, **12**, 151
6 J. Simeray, J. P. Chaumont, F. Bevalot and I. Vaquette, *Fitoterapia*, 1982, **53**, 48.
7 I. Chinou, C. Demetzos, C. Harvala, C. Roussakis and J.F. Verbist, *Planta Med.*, 1994, **60**, 36.
8 E. Yesilada, O. Ustun, E. Sezik, Y. Takaishi, Y. Ono and G. Honda, *J. Ethnopharmacol.*, 1997, **58**, 73.
9 H. Bouamama, J. Villard, A. Benharref and M. Jana, *Therapie*, 1999, **54,** 733.
10 H. Mekhfia, M. El Haouari, A. Legssyer, M. Bnouhama, M. Aziz, F. Atmani, A. Remmal and A. Ziyyat, *J. Ethnopharmacol.*, 2004, **94**, 322.
11 M. Aziz, N. Tab, A. Karim, H. Mekhfi, M. Bnouham, A. Ziyyat, A. Melhaoui and A. Legssyer, *Fitoterapia*, 2006, **77**, 428.
12 A. Danne, F. Petereit and A. Nahrsted, *Phytochem.*, 1994, **37**, 538.
13 G. J. Vernin, *J. Essent. Oil Res*, 1993, **5**, 563.
14 N. Chaves, J.J. Rios, C. Gutierrez, J.C. Escudero and J.M. Olías, *J. Chromatogr. A*, 1998, **799**, 115
15 T. Sosa, N. Chaves, J.C. Alias, J.C. Escudero, F. Henao and C.J. Gutiérrez-Merino, *Chem. Ecol.*, 2004, **30**, 1101
16 T. Sosa, J. C. Alias, J. C. Escudero and N. Chaves, *Biochem. Syst. Ecol.*, 2005, **33**, 364
17 D. Andrade, C. Gil, L. Breitenfeld, F. Dominguesa and D.A. Paula, *Ind. Crops Prod.*, 2009, **30**, 167
18 AOCS, *Champaign: American Oil Chemists' Society*, 1990.
19 S. Sakanaka, Y. Tachibana and Y. Okada, *Food Chem.*, 2005, **89**, 575.
20 U.R. Kuppusamy, M. Indran and B.R.S. Balraj, *J. Trop. Med. Plants*, 2002, **3**, 5

ADDITION OF LEMON AND ORANGE FIBERS AS FUNCTIONAL INGREDIENTS TO A SWEETENED CREAM CHEESE

L. Trigueros, E. Sendra, E. Sayas-Barberá, C. Navarro, J.A. Pérez-Álvarez and J. Fernández-López

IPOA Research Group (UMH-1 and REVIV_ Generalitat Valenciana) , Food Technology Department, Universidad Miguel Hernández, Ctra. de Beniel km 3.2, 03312, Orihuela (Alicante) SPAIN.

1 INTRODUCTION

Citrus fruits are good sources of bioactive compounds (mainly polyphenols) and dietary fiber (DF).[1] Fiber powders obtained from citrus juice industry by-products are good sources of bioactive compounds and functional properties (water and oil binding capacity).[2]

The recommended daily intake of DF in Mediterranean countries is about 20 g/d for men and 15,7 g/d for women.[3] Citrus by-products constitute an important source of fiber since they are very rich in pectins.[4]

Fiber enriched dairy products are perceived by consumers as healthy foods. Fermented dairy products are also perceived as healthy foods and fruit enriched or flavoured dairy products have good acceptance among consumers.

Cream cheese is a fresh cheese made using a mesophilic lactic acid starter culture. Fibers may interact with the microbial populations of cream cheese; either residual essential oils may inhibit their growth or fiber components as oligosaccharides may enhance their growth.[5] The sensory impact of functional ingredients incorporated into cheeses has not been very extensively investigated. However, it has been assumed that products containing these ingredients have different taste profiles, when compared to conventional, non-functional products.[6]

Cream cheese is a versatile food that permits addition of other ingredients including fibers like citrus fiber. The production of a cream cheese capable of generating a potentially functional effect, due to the incorporation of DF is promising. The aim of the present study was to develop a sweetened cream cheese enriched with lemon and orange fibers and to characterize their sensory characteristics and establish its shelf-life.

2 METHOD AND RESULTS

2.1 Experimental design

Six pilot-scale cheese-making trials were performed. Three of them for the physicochemical and shelf life study under two storage temperatures (5 and 25 °C) and the other three for sensory analysis.

2.2. Cream cheese manufacture

Three types of cream cheese with (i) 0%, (ii) 1% lemon fiber, (iii) 1% orange fiber were manufactured using *Lactococcus lactis* subsp. *lactis*, *Lactococcus lactis* subsp. *cremoris*, *Lactococcus lactis* subsp. *lactis* biovar *diacetylactis* and *Streptococcus thermophilus* (EZAL® MA400; Rhodia Food, Sassenage, France) as starter cultures. Fiber was obtained from lemon and orange juice by-products by a procedure described by Lario et al. (2004).[7]

Cheese-base was manufactured from pasteurized and homogenized milk (3.6% fat) with 8% sugar and 1% salt and cheese-fiber was supplemented with 1% fiber. The mixture was pasteurized at 80 °C for 30 minutes, afterwards it was cooled to 35 °C and starter cultures were added. Coagulation occurred in 6 hours when the pH of the mixture reached 4.6. Cream (13.33%) was added and the mixture was pasteurized again at 65 °C for 15 minutes. The curd was placed in sterilized cotton cheese-clothes, and allowed to drain by gravity at room temperature for 3 hours.

Once cream cheese was obtained it was distributed in sterile plastic flasks (100 mL volume). Half of the flasks were stored at 5 °C and the other half at 25 °C to evaluate shelf-life by microbial counts.

2.3. Physico-chemical and compositional analysis

Cheese yield was calculated (kg cheese/100 L milk). Moisture, dry matter, fat, ash, acidity, calcium and protein contents were determined for cream cheese and whey one day after cheese making by AOAC methods.[8] The pH values of cheeses were determined. Syneresis was measured at 7 °C by drainage of 20 g aliquots through a sieve of 120 mesh under vacuum during 10 minutes.

Cheese yield and syneresis of day 1 are shown in Table 1. No significant differences were detected in cheese yield ($P > 0.05$) but cheeses with citrus fiber had higher yield than control cheese, probably due to the high water holding capacity (WHC) of the citrus fiber: 10.59 g water/g fiber[9] and 11.8 g water/g fiber[7] for orange and lemon fiber respectively. Citrus fiber addition significantly ($P < 0.05$) decreased syneresis, which is closely related to the WHC of the fiber, so whey is retained in those cheeses that contain fiber[10] and this is a positive effect because separation of whey is a fault when it occurs on the surface of the cheese.[11] No significant differences ($P > 0.05$) were detected in chemical composition and pH values of cheeses (Table 1).

Table 1 *Cheese yield, syneresis* and *composition of cream cheese with lemon and orange fibers.*

Cheese	*Y*	*S*	*M*	*C.P*	*F*	*A*	*C*	*L.A.*	*pH*
Control	61.44^a	49.67^b	75.97^a	4.86^a	7.92^a	0.78^a	0.114^a	0.74^a	$4.60^a\pm$
	±4.05*	±2.67	±1.50*	±0.47	±0.25	±0.02	±0.06	±0.54	0.07
Lemon	68.67^a	45.43^a	77.53^a	4.70^a	8.75^a	0.84^a	0.115^a	0.70^a	$4.50^a\pm$
	±3.06	±1.48	±0.59	±0.27	±0.78	±0.02	±0.03	±0.24	0.06
Orange	71.33^a	42.68^a	76.94^a	4.45^a	8.25^a	0.80^a	0.113^a	0.67^a	$4.54^a\pm$
	±3.86	±2.13	±0.58	±0.67	±0.75	±0.02	±0.07	±0.32	0.05

* Mean values ± Standard Error (n = 9). Mean values with different superscript letters in a same column denote significant differences ($P < 0.05$).
Abbreviations: Y=Cheese Yield (%); S=Syneresis (%); M=Moisture (%); C.P.=Crude Protein (%); F=Fat (%); A=Ash(%); C= Calcium (mg/100g); L.A.=Lactic Acid(%).

The addition of fiber did not significantly affect whey composition ($P > 0.05$). Calcium and lactic acid contents were higher in whey than in cheese (Table 2), due to the solubilization of micellar calcium provoked the low pH of the cheese.

Table 2 *Composition of whey of cream cheese with lemon and orange fibers.*

Type of whey	*Dry matter (%)*	*Crude Protein (%)*	*Ash (%)*	*Calcium (mg/100g)*	*Lactic acid (%)*
Control	$14.54^a \pm 0.59$*	$0.44^a \pm 0.01$	$0.78^a \pm 0.05$	$0.133^a \pm 0.43$	$3.50^a \pm 0.06$
Lemon	$13.89^a \pm 0.20$	$0.39^a \pm 0.10$	$0.83^a \pm 0.01$	$0.133^a \pm 0.13$	$3.69^a \pm 0.11$
Orange	$13.72^a \pm 0.25$	$0.35^a \pm 0.01$	$0.78^a \pm 0.01$	$0.124^a \pm 0.01$	$3.49^a \pm 0.01$

* Mean values ± Standard Error (n = 9). Mean values with different superscript letters in a same column denote significant differences ($P < 0.05$).

2.4. Sensory analysis

Three different tastings sessions were run to evaluate product acceptability, select fiber size and establish shelf life. Panelists (36) were members of the staff and students of the Miguel Hernández University, Alicante, Spain. All sensory work was carried out in the sensory laboratory at the University, which fulfills requirements according to the International Standards (ISO, 1988). During the evaluation, the panelists were situated in private booths under incandescent light. Tap water was provided between samples to cleanse the palate. The sample presentation order was randomized for each panelist. Several commercial flavourings (lemon, orange, and strawberry), colorings (red) and fiber sizes (< 0.4 mm, and 0.4 < x < 0.7 mm) were tested to evaluate product acceptability in a seven-point hedonic scale. A seven-point scale (from imperceptible to extremely high) was used to rate the following parameters: brightness, yeasty smell and acidity. Samples were scored for overall acceptability during three weeks (weekly testings) of storage at 5 °C to determine cream cheese shelf life in a seven-point hedonic scale.

The objective of most quantitative consumer research conducted in support of product development is to determine consumers' affective reaction to new or revised products.[12] In this study we wanted to know whether masking the product with flavourings or colorings alters the perception of tasters. Sensory evaluation results of average scores from 4.44 to 5.68, with no significant differences among them ($P > 0.05$). However, cream cheeses without fiber were the best scored. In general, cheeses without flavourings and colorings were preferred over those that contained them.

Table 3 *Sensory scores in a 7-point hedonic scale for cream cheese during three weeks of refrigerated storage at 5 °C.*

Type of cheese	*Overall acceptability*		
	1st week	*2nd week*	*3rd week*
Control	$5.22^a \pm 0.19$*	$5.59^a \pm 0.18$	$5.13^a \pm 0.25$
Lemon	$5.13^a \pm 0.21$	$5.33^a \pm 0.23$	$4.48^a \pm 0.29$
Orange	$5.43^a \pm 0.29$	$4.96^a \pm 0.28$	**

* Mean values ± Standard Error (n = 9). Mean values with different superscript letters in a same column denote significant differences ($P < 0.05$).
** Not tested for deterioration in the sample.

Fiber sizes were tested and average results ranged from 5.07 to 5.7, showing a non significant tendency ($P > 0.05$) to prefer a fine fiber size, so this fiber size (< 0.4 mm) was used for the manufacture of cream cheeses at larger scale.

The end of shelf-life of foods is ultimately assessed by sensory failure. The deterioration in sensory quality is quantified by monitoring changes in specific attributes or reduction in overall quality/acceptability.[13] In the present work overall acceptability was measured during three weeks (Table 3) and no statistical differences between samples were found but all types of cream cheeses received average scores of at least 4.48 on a 7-point scale.

In several studies the use of different cut-off points to mark end of shelf-life of food had been used.[14,15] On a 7-point scale, an off-flavor score of 2.5 was used for one product while score of 3.5 was used for another.[16] In our study we measured the specific sensory attribute 'yeasty smell' after seven days of refrigerated storage and we assessed cheeses scored a maximum of 1.70 in a 7-point hedonic scale (Figure 1). Also, there was a tendency (not significant $P > 0.05$) to increase acidity and brightness perception in cheeses with citrus fiber.

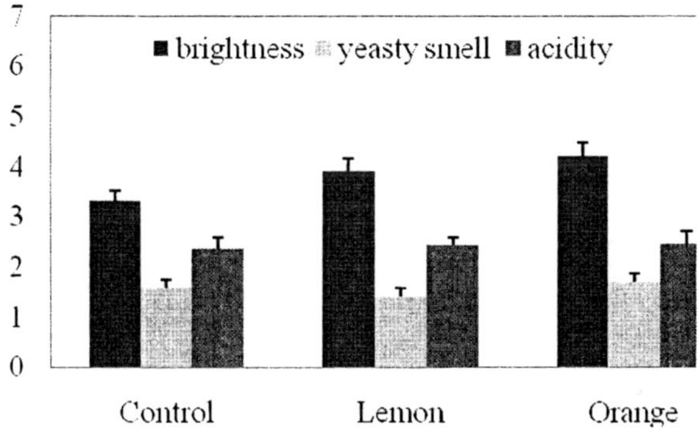

Figure 1 *Sensory scores for brightness, yeasty smell and acidity (in a 7-point scale) obtained by cream cheese at 7 days of refrigerated storage (5 °C). Mean values ± Standard Error (n = 9).*

2.5. Microbiological analysis

Microbial analysis of cheese included determinations of the number of total aerobic mesophilic bacteria (AMB), counts of enterobacteria, moulds and yeasts and lactic acid bacteria (LAB) during shelf-life period. Media used were: LAB in Man Rogosa Sharp (MRS) agar incubated in anaerobic conditions for 72 h at 37 °C. AMB, enterobacteria and moulds and yeasts were determined by plating the diluted samples in 3M Petrifilm™ (3M España S.A., Madrid, Spain) AMB plates were incubated at 35 °C for 48 h, enterobacteria at 37 °C for 24 h and moulds and yeasts at 25 °C for 5 days.

Changes in microbial counts and pH values during storage are shown in Figure 2. AMB, LAB and moulds and yeasts counts were not significantly affected by the presence of fiber.

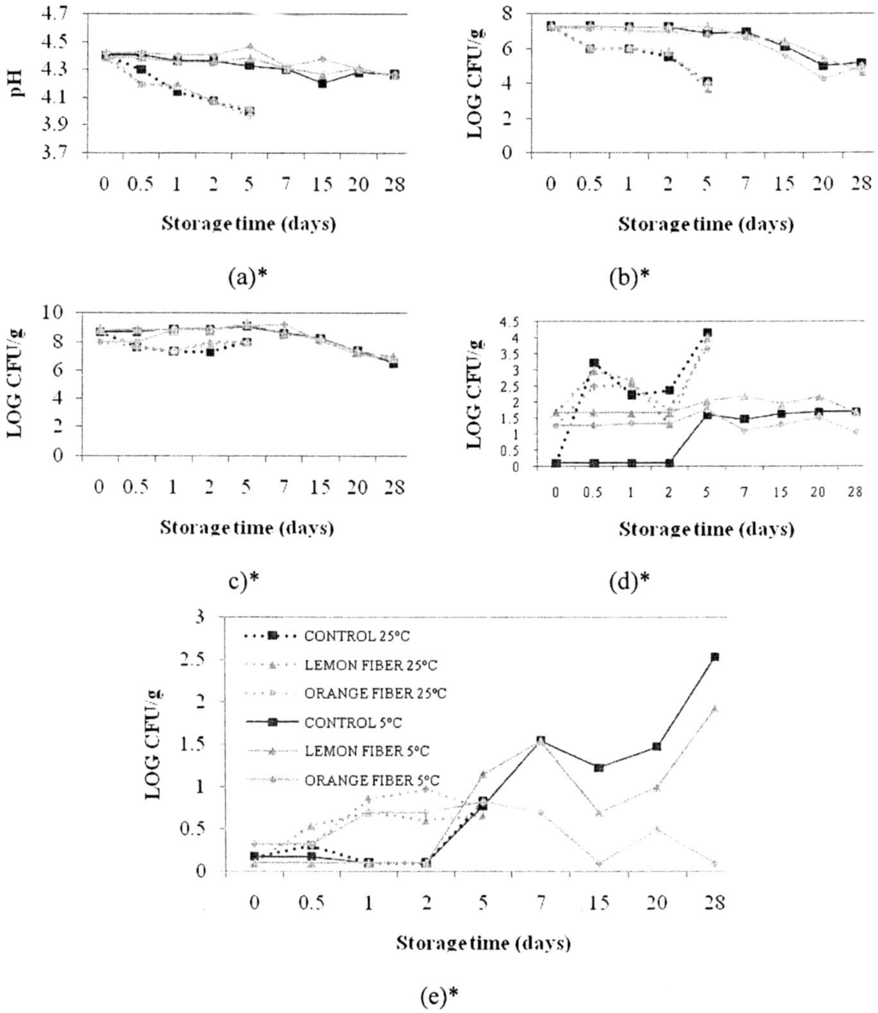

Figure 2 *Evolution of pH (a) and counts of lactic acid bacteria (b), aerobic mesophilic bacteria (c), moulds and yeasts (d) and enterobacteria (e) in cream cheese with citrus fibers stored at 5 °C and 25 °C for 28 days (n = 9). *Legend is the same for all the figures.*

The number of LAB in samples stored at 5 °C were constants during the first 5 days of storage and decrease thereafter. Al-Kadamany et al. (2003)[17] reported a similar behavior of *labneh* at same temperature but they obtained lower counts. At 25 °C counts decreased during storage. The sharp drop in LAB counts is due to the inhibitory effects of relatively high temperatures and increased acidity on the growth of yogurt cultures.[18] Moulds and yeasts growth limited the shelf-life of cream cheese stored at 25 °C, whereas enterobacteria were the limiting population for cream cheese stored at 5 °C. The presence of orange fiber decreased counts of enterobacteria. The presence of enterobacteria was chosen to mark the end of shelf-life (Figure 2).

3 CONCLUSION

The determined shelf-life of cream cheese was 28 days at 5 °C judging the growth of enterobacteria. Orange and lemon fibers are promising ingredients for the preparation of cream cheeses, their presence increased cheese yield and did not modify overall composition or microbial quality of the cheese. Further studies are needed regarding sensory properties.

References

1 S. Gorinstein, O. Martín-Belloso, Y.S. Park, R. Haruenkit, A. Lojek, M. Ciz, A. Caspi, I. Libman and S. Trakhtenberg, *Food Chem*, 2001, **74**, 309.
2 L. Aleson-Carbonell, J. Fernández-López, J.A. Pérez-Álvarez and V. Kuri, *In. Food Sci. and Emer. Technol.*, 2005, **6**, 247.
3 R. Capita and C. Alonso-Calleja, *J Food Sci Nutr*, 2003, **54**, 57.
4 A. Askar, *Fruit in Processing*, 1998, **8**, 273.
5 E. Sendra, P. Fayos, Y. Lario, J. Fernández-López, E. Sayas-Barberá and J.A. Pérez-Álvarez, *Food Microbiol*, 2008, **25**, 13.
6 H.R. Cardarelli, F.C.A. Buriti, I.A. Castro and S.M.I. Saad, *LWT Food Sci. Technol.*, 2008, **41**, 1037.
7 Y. Lario, E. Sendra, F.J. García-Pérez, E. Sayas-Barberá, J. Fernández-López and J.A. Pérez-Álvarez, *In. Food Sci. Emer. Technol.*, 2004, **5**, 113.
8 AOACS in *Official Methods and Recommended Practices of the American Oil Chemists' Society*, 4th edn, American Oil Chemists' Society, Champaign, 1990.
9 F.J. García, E. Sendra, J.M. Fernández, J. Fernández-López, E. Sayas and J.A. Pérez-Álvarez, *Alimentaria*, 2003, 71.
10 E. Sendra, Y. Lario, J.A. Pérez-Álvarez, and J. García, *Alimentación, Equipos y Tecnología*, 2004, **186**, 49.
11 J.A. Lucey, *Food Hydrocolloid*, 2001, **15**, 603.
12 H.R. Cardarelli, F.C.A. Buriti, I.A. Castro and S.M.I. Saad, *LWT Food Sci. Technol.*, 2008, **41**, 1037.
13 E. Al-Kadamany, I. Toufeili, M. Khattar, Y. Abou-Jawdeh, S. Harakeh and T. Haddad, *J. Dairy Sci.*, 2002, **85**, 1023.
14 R.M.V. Alves, C.I.G.L. Sarantopulos, E.S. Saron and M.R. Bordin, *Packag. Sci. Technol.*, 2001, **14**, 79.
15 C.W. Fritsch, C.N. Hofland and Z.M. Vickers, *J. Food Sci.*, 1997, **62**, 425.
16 M.C. Gacula Jr, *J. Food Sci.*, 1975, **40**, 400.
17 E. Al-Kadamany, M. Khattar, T. Haddad and I. Toufeili, *LWT Food Sci. Technol.*, 2003, **36**, 407.
18 H. Rohm, F. Lechner and M. Lehner, *J. Food Protect.*, 1990, **51**, 478.

APPLICATION OF ORANGE FIBRE AS A FUNCTIONAL INGREDIENT IN botifarró: A SPANISH BLOOD-SAUSAGE

E. Sayas-Barbera, E. Sendra, C. Navarro, J.A. Pérez-Alvarez, E. Sánchez-Zapata, M. Viuda-Martos and J. Fernández-López

Agro-Food Technology Department. Universidad Miguel Hernandez. Escuela Politécnica Superior de Orihuela. Crta. Beniel km 3,2 (03312) Orihuela, Alicante, Spain. IPOA Research Group (UMH-1 Group REVIV, Generalitat Valenciana).

1 INTRODUCTION

botifarró is one of the most traditional sausages from the Baleares Islands (Spain). It has an ancient origin; it is an evolution of meat products introduced by the Romans. Studies on *botifarró* are very limited, as its elaboration is linked to traditional slaughter and to very specific rural areas. *Botifarró* is a cooked pork meat sausage of approximately 50-100g of weight, cylindrical shape (2 to 4 cm of diameter) and 10 to 15 cm of length. It has a characteristic dark colour, due to the blend of pork meat with blood. It is made from a mixture of lean meat, animal fat, blood, salt and different spices such as black pepper, paprika, anisette and pinenuts. This mixture is stuffed in natural casings and the product is boiled for 1h at 90-95°C: Optionally it could carry a maturation stage for two months.[1]

Increasing concerns about the potential health risk associated with the consumption of high-fat foods has led the food industry to develop new formulations or modify traditional products to make them healthier.[2] These modifications include the use of non-meat ingredients that introduced certain functional characteristics. In this regard, dietary fibre (DF) have been successfully used in meat products processing in both dry-cured and cooked sausages. [3,4,5,6,7,8]

Recent studies suggest that DF acts as a protective agent against cardiovascular diseases, diverticulosis, constipation, obesity, artherosclerosis, irritable colon, colon cancer and diabetes. [9,10] These physiological functions are attributed to two groups of compounds from DF: water-soluble (pectins and gums) and those water insoluble (cellulose, hemicellulose and lignin) polysacharides.[10] It is widely accepted that citrus fibre extracts have better quality than other sources of dietary fibres, probably due to the presence of associated bioactive compounds with antioxidant properties.[7,10,11] Whereby, *botifarró* formulated with orange fibre could have a positive effect on consumer heath and it could improve their sensory characteristics to increase the potential market of the product and to satisfy consumer demands.

The main aim of this work was to study the effect of the addition of orange fibre (from citrus juice industry by-products), as a potential functional ingredient, on the quality characteristics of *botifarró* (Spanish blood sausage).

2 MATERIALS AND METHODS

2.1 Sausage manufacture

Orange fibre was obtained directly from orange juice by-products, following the optimized procedure described by Fernández-Ginés et al.[4] to obtain a powder with a diameter <0,5 mm. Sausages were manufactured in IPOA Research group pilot plant according to industrial processing. Traditional formula of *botifarró* was used: 50% pork lean, 37,5% pork backfat 12% blood (adding up to 100%, the other ingredients were calculated based on them), 1% salt, 0.5% phosphates, 0.03 % sodium nitrite and a mixture of different spices that included paprika, black pepper, clove and a commercial mixture for "*morcilla*".

The meat and fat cuts were minced through a 8 mm plate (Cato, Sabadell, Spain), and mixed with other ingredients. This original mixture was split into 3 batches to which orange fibre was added in different amounts (0, 0.5%, 1%, 1.5% and 2%). The mixtures were stuffed into natural casings (pork thin intestine) and transferred to a cooking container and boiled in water at 90°C for around 1h. Three independent batches were manufactured on the same day. Thus, the values shown in the tables and figures are the mean of these nine separate trials. Chemical composition and physico-chemical characteristics were determined, as well as sensory properties.

2.2 Chemical analysis

Moisture (oven air-drying method), ash (muffle furnace), protein (Kjeldhal nitrogen) and fat content (Soxhlet method) were determined following AOAC procedures.[12] Residual nitrite level (mg $NaNO_2$/kg sample) was determined according to the standard method ISO/DIS 2918.[13]

2.3 Physico-chemical analysis

The CIE LAB colour space was studied and the following colour coordinates were determined: lightness *(L*)*, *a** coordinate (redness/greenness, ±) and b* coordinate (yellowness/bluesness, ±). All colour measurements were made according to AMSA Guidelines.[14] Colour determinations were made by means of a Minolta CM-2600D (Minolta Camera Co., Osaka, Japan) spectrophotometer with illuminant D_{65} and 10° as observer. Spectrally pure glass (CR-A51; Minolta Co.) was placed between the samples and the equipment.

Texture profile analysis (TPA) was performed with a Texture Analyser TA-XT2 (Stable Micro Systems, Surrey, England). Sausage samples were removed from casing, cut into slices (1cm thick) and subjected to a 2-cycle compression at 7°C. The texture profile parameters (hardness (g), cohesiveness, springiness (mm) and chewiness (gxmm) were obtained.

Water activity (Aw) was measured at 25°C by using an electric hygrometer NOVASINA TH200 (Novasina; Axair LTD., Pfaeffikon, Switzerland). The pH was measured in a homogenate prepared with 15g sausages and 150mL of deionised water, using a Crison pH meter (Model 507).

2.4 Sensory evaluation

Untrained panellists (30), consumers of this type of products, were recruited from the staff and students of the Miguel Hernandez Univ. (Alicante, Spain). During evaluation, each panellist evaluated 5 slices (approximately 1.5 cm thick) corresponding to: sample control and sample with 0.5, 1, 1.5 and 2 % of added fibre. The attributes were measured on a 8-point scale and their descriptors were as follow: colour (from 1-extremely light to 8-extremely dark), taste (from 1-extremely bitter to 8-extremely acid), hardness (from 1-extremely soft to 8-extremely hard), granularity (from 1-absence of granules to 8-extremely gritty), juiciness (from 1-extremely dry to 8-extremely moist), cohesiveness (from 1-extremely crumbly to 8-extremely cohesive), fattiness (from 1-extremely lean to 8-extremely fatty), saltiness (from 1-imperceptible to 8-extremely salty), odour (from 1-imperceptible to 8-extremely intense) and overall quality (from 1-extremely dislike to 8-extremely like). All sensory work was carried out in the sensory laboratory at the University, which fulfils requirements according to the international standards.[14,15] During evaluation the panellists were situated in private booths under incandescent light, with an intensity of approximately 350 lux.

2.5 Statistical analysis

All data were statistically analysed using ANOVA procedures. Tukey's test for multiple comparisons was performed using the general linear model of SPSS 16.0 for Windows (SPSS, Chicago, IL, USA) software package. The level of statistical significance was $P < 0.05$.

3 RESULTS AND DISCUSSION

Table 1 shows the results of the chemical analysis of the 5 different *botifarró* formulations. There were no significant differences ($P<0.05$) in main components (moisture, protein and lipids) for the addition of fibre. Ash content was higher ($P<0.05$) in blood sausages with 2% fibre. The increase of ash content was related to the concentration of added fibre.

It is very important to note that the addition of orange fibre (all concentrations) reduces the residual nitrite level: This behaviour is related to fibre copassengers': bioactive polyphenols.[4] This decrease in residual nitrite has healthy effects because it reduces the possibility of nitrosamines formation, which is considered a risk associated to the consumption of meat products containing nitrite. This reduction in residual nitrite has been previously reported by other authors, in other cooked products (bolognas) and dry-cured sausages enriched with citrus fibres.[4,5]

Table 1 *Chemical composition, pH and a_w of blood sausages (botifarró) formulated with different citrus fibre concentrations (mean ± standard deviation)*[a].

		Orange fibre			
	control	0.5 %	1.0 %	1.5 %	2 %
Moisture (%)	50.74±1.13a	50.51±6.26a	55.67±4.97a	50.92±3.66a	53.49±1.39a
Protein %	12.67±0.16a	13.19±0.11a	11.93±0.26a	13.03±1.04a	12.30±0.37a
Lipids %	33.83±0.18a	34.82±2.21a	31.13±2.50b	33.81±2.24a	29.17±2.32b
Ash %	0.80±0.15a	1.41±0.28b	1.49±0.16b	1.60 0.02b	1.99±0.01c
Nitrite (mg/kg)	40.28±3.16a	32.92±3.02a	12.31±2.36 b	10.08±0.72b	7.55±0.30b
pH	6.53 ±0.50a	6.64±0.47a	6.41±0.02a	6.44±0.61a	6.38±0.45a
Aw	0.961±0.001a	0.966±0.001a	0.963±0.001a	0.964±0.001a	0.962±0.001a

[a] Means followed by a different letters within the same row are significantly different ($P<0.05$).

Moreover, *botifarró* is a perishable product due to its rich nutrients composition, high pH (up 6.3) and high water activity 0.96 (table 1), as other cooked meats containing blood.[16] The addition of fibre didn't affect pH and a_w. It has already been reported in other studies that citrus fibre addition produced a pH decrease due to the presence of organic acids and others acid compounds in orange fibre.[17] In the present study the incorporation of fibre may have decreased meat pH but the cooking process compensated this effect due to the release of buffering compounds and fat during cooking.[18]

Lightness, redness and yellowness showed differences ($P<0.05$) due to orange fibre concentration (table 2). The behaviour of L* and a* differed between high and low fibre concentrations. L* values increased with the addition of 0.5 and 1% citrus fibre, while at higher concentrations of fibre, the effect was just the opposite. Fernández-Ginés et al.[18] attributed the increase in L* values to the fibre incorporation, due to its high content of white components (albedo). The fact that increased orange fibre concentrations (concentration up 1.5%) reduced product lightness could be explained by the masking effect of the emulsified matrix containing blood; together with the increase in water holding capacity which reduces free water and decreases light reflection. Regarding redness (a*) the results (table 2) showed differences ($P<0.05$) for orange fibre concentration: the presence of fibre decreased a* values at low fibre concentration (0.5%, 1% y 1.5%). In samples with high fibre concentration (2%), orange fibre addition did not modify a* values of *botifarró*. This recovery or maintenance of a* values observed at 2 % fibre enriched *botifarró* may be due to the presence of reducing substances (polyphenoles, ascorbic acid, etc.) in the fibre,[10] which favours the formation and stabilization of nitrosomyoglobin by reducing compounds (polyphenol and ascorbic acid) that exists in citrus fibres.[2,4] In other meat products small amounts of citrus fiber increased redness, but in this product, the high amount of blood reduces its shelflife, by liberation of nonheme iron, which is related to meat oxidations. Thus orange fibre at low concentrations didn't protect *botifarró* from oxidations. The increase of yellowness with added fibre ($P<0,05$), could be attributed to the carotenes from flavedo retained in the orange fibre. These compounds would not be removed in the washing steps.[4,18] Other authors have reported colour changes induced by the addition of fruit or cereal fibres to different meat products.[19,20,21,22]

The addition of citrus fibre modified the textural properties of blood sausages. All the parameters followed the same behaviour: at low concentrations of fibre the values increased. However, at high concentrations the effect was inverse, the values decreased.

Table 2 *Colour (CIE L*, a*, b*) and texture profile analysis of blood sausages (botifarró) formulated with different citrus fibre concentration (mean ± standard deviation)[a].*

		Orange fibre			
	control	0.5 %	1.0 %	1.5 %	2 %
Colour					
L*	37.14±0.80c	43.27±6.30b	45.07±4.86a	38.95±0.93c	37.98±2.24c
a*	8.59±0.59b	7.89±1.97c	6.88±1.79d	7.98±0.95c	9.04±0.70a
b*	6.66±0.37c	6.11±1.10c	7.49±1.29b	7.11±1.44bc	8.09±1.04a
Texture parameters					
Hardness (g)	7130±339b	15795±577a	10087±487ab	2609±628c	988±61c
Cohesiviness	0.30±0.23b	0.50±0.03a	0.36±0.21b	0.32±0.13b	0.18±0.21c
Springiness (mm)	4.11±0.30a	4.33±0.11a	4.32±0.21a	3.60±0.74b	4.07±0.32a
Chewiness (gxmm)	11164±120b	33567±105a	19225±922b	3666±324c	580±72c

[a] Means followed by a different letters within the same row are significantly different ($P<0.05$).

Table 3 *Effect of addition of orange fibre on the sensory properties of blood-sausages (botifarró) (mean ± standard deviation)*[a].

		Orange fibre			
	control	0.5 %	1.0 %	1.5 %	2 %
Colour	4.18±1.06ab	4.50±1.10a	4.20±1.06ab	3.60±1.27b	3.80±0.89b
Taste	4.78 ±0.80a	4.65±0.88a	4.55±0.69a	4.45±1.36a	4.70±0.73a
Hardness	4.18±1.20ab	4.95±1.39a	4.45±1.19ab	4.35±1.21a	3.70±1.08b
Granularity	4.95±1.26a	4.55±1.15a	5.45±0.69a	5.15±1.04a	4.95±1.36a
Juiciness	5.30±1.20b	5.10±1.29b	4.90±1.29c	4.10±1.29d	5.60±1.23a
Cohesiveness	5.20±0.97a	5.70±0.8a	5.35±0.93a	5.35±1.46a	4.50±0.83b
Fattiness	5.55±1.01b	5.85±0.99a	4.75±1.37c	4.20±1.32c	5.35±1.14b
Saltiness	4.40±0.93a	4.60±0.75a	4.15±0.81a	4.45±1.00a	4.00±0.92a
Odour	4.63±1.03a	4.60±1.10a	4.85±0.88a	4.45±1.15a	4.65±0.93a
General quality	4.45±1.26b	5.00±1.21a	4.75±0.91ab	4.20±1.44b	4.60±1.27ab

[a] Means followed by a different letters within the same row are significantly different ($P<0.05$).

This behaviour could be attributed to the characteristics of the matrix that is formed: fibre addition could modify the ultrastructure of the meat products, and other properties such as ionic strength of the medium.[4,23] When sausages were manufactured with low concentrations of added fibre (0.5%) all textural parameters significantly ($p<0.05$) increased (table 2). However, the addition of orange fibre (up 1.5%) to the mixture could disrupt the protein-water or protein-protein gel network which, in turn, could decrease the gel strength of the products.[21] *Botifarro* with 2% added fibre had markedly decreased textural properties, with the exception of springiness.

The results from sensory evaluation are showed in table 3. Colour, hardness, juiciness, cohesiveness, fattiness and general quality were affected ($P<0.5$) by fibre addition. The sausages with 0.5 and 1% de fibre had a darker colour, this results are well correlated with the values of lightness (L*) obtained by instrumental methods (table 2). Hardness and cohesiveness differences detected by Texture Analyzer were detected by panellist; the addition of orange fibre at small concentrations (0.5-1.5%) increased hardness and cohesiveness perception; whereas sausages with 2% fibre received smaller scores for hardness and cohesiveness. As explained before, this behaviour (at high and low fibre concentrations) could be related to interactions between proteins and fibre,[2] the addition of up to 1.5% could reduced the stability and strength of the meat matrix. Some authors reported a hardening effect when fibre is added to various meat products.[4] *Botifarró* with 2% added orange fibre was perceived to have high juiciness, with no differences on fatness when compared to control, low saltiness and with light colour. Regarding overall quality, the panellist preferred the sausages with 0.5% fibre to the traditional ones (control).

4 CONCLUSION

Orange fibre addition modifies significantly the physico-chemical and sensory characteristics of *botifarró*. The changes in all parameter under study depend on the orange fibre dose. These findings can have significant implications for processors and may offer them the opportunity to improve the nutritional and heath qualities of their products and may have beneficial effects (including the decrease in residual nitrite levels). The best sensory scores were obtained for sausages with orange fibre (0.5, 1 and 2%), additionally 2% enriched *botifarró* have lowest nitrite residual level.

References

1 J.M. Ginés-Fernández, J.A. Pérez-Álvarez, J. Fernández-López, M.E. Sayas-Barberá, C. Navarro, Y. Lario, E. Sendra and J. Martorell, *Eurocarne*, 2006, **149**.
2 L. Aleson-Carbonell, J. Fernández-López, M.E. Sayas-Barberá, E. Sendra and J.A. Pérez-Álvarez, *J. Food Sci. Agric.*, 2004, **84**, 2077.
3 M.L. García, R. Domingues, M.D. Galvez, C. Casas and M.D. Selgas, *Meat Sci.*, 2002, **60**, 227.
4 J.M. Fernandez-Ginés, J. Fernández-López, M.E. Sayas-Barberá, E. Sendra and J.A. Pérez-Álvarez, *J. Food Sci.*, 2003, **68**, 710.
5 L. Aleson-Carbonell, J. Fernández-López, E. Sendra M.E. Sayas-Barberá and J.A. Pérez-Álvarez, *J. Sci. Food Agric.*, 2004, **84**, 2077.
6 E. Caceres, M.L, García, J. Toro and M.D. Selgas, *Meat Sci.*, 2004, **68**, 87.
7 J. Fernández-López, J.M. Fernandez-Gines, L. Aleson-Carbonell, E. Sendra, M.E. Sayas-Barberá, and J.A. Pérez-Álvarez, *Trends Food Sci. Technol.*, 2004, **15**, 176.
8 V.S. Eim, S. Simal, C. Rosselló and A. Femenia, *Meat Sci.*, 2008, **80**, 173.
9 R. Rodríguez., A. Jiménez, J. Fernández-Bolaños, R. Guillén and A. Heredia, *Trends Food Sci. Technol.*, 2006, **17**, 3.
10 J. Fernández-López, E. Sendra, C. Navarro, M.E Sayas-Barberá, M. Viuda-Martos and J.A. Pérez-Álvarez, *Food Sci. Technol.*, 2009, **44**, 748.
11 Y. Lario, E. Sendra, J. García-Pérez, C. Fuentes, E. Sayas-Barberá, J. Fernández-López and J. A. Pérez-Alvarez, *Innovative Food Sci. Emerging Technol.*, 2004, **5**, 113.
12 *Official methods of analysis,* AOAC, (16th ed.) Association of Official Analytical Chemists, Washington. DC: 2005.
13 *International Standard 2918*. Meat and meat products. ISO 2918:1975.
14 *Guidelines for meat color evaluation*. AMSA. American Meat Board, Chicago IL, 1991.
15 *Guidelines for sensory evaluation laboratories*. STP 913. ASTM. American Society for Testing and Materials, Philadelphia, 1986.
16 E.M. Santos, C. González-Fernández, I. Jaime and J, Rovira, *Meat Sci.*, 2003, **65**, 893.
17 J. Fernandez-Lopez, M. Viuda-Martos, E. Sendra, E. Sayas-Barberá, C. Navarro and J.A. Pérez-Álvarez, *Eur. Food Res. Technol.*, 2007, **226**,1.
18 J.M. Fernandez-Ginés, J. Fernandez-Lopez, E. Sayas-Barberá, E. Sendra, and J.A. Pérez-Álvarez, *Meat Sci.*, 2004, **67**, 7.
19 J.R. Claus and M.C. Hunt, *J. Food Sci.*, 1991, **56**, 643.
20 E.S. Troutt, M.C. Hunt, D.E. Johnson, J.R. Claus, C.L. Kastner and D.F. Kropf, *J. Food Sci.*, 1992, **57**, 19.
21 N. Grigelmo-Miguel, M.I. Abadía-Serós and O.A. Martín-Belloso, *Meat Sci.*, 1999, **52**, 247.
22 S. Cofrades, M.A. Guerra, J. Carballo, F. Férnandez-Martín and F. Jiménez-Colmenero, *J. Food Sci.*, 2000, **65**, 281.
23 J.R. Claus, M.C. Hunt, C.L. Kastner and D.H. Kropf, *J. Food Sci.*,1990, **55**, 338.

PREBIOTIC POTENTIAL AND ANTIMICROBIAL EFFECT FROM A BY-PRODUCT OF THE ALMOND PROCESSING INDUSTRY

G. Mandalari,[1] A. Tomaino,[2] G. Bisignano,[2] A. Narbad,[3] K.W. Waldron[4] and M.S.J. Wickham[1]

[1]Model Gut Platform, [3]Integrated Biology GI Tract Programme, and [4]Sustainability of the Food Chain Platform, Institute of Food Research, Norwich Research Park, Colney, Norwich NR4 7UA, UK
[2]Pharmaco-Biological Department, University of Messina, Messina, 98168, Italy

1 INTRODUCTION

The global production of almonds is around 1.7 million metric tons, with California currently producing 80% of the world's almonds.[1] Positive effects of almond consumption have been demonstrated in reducing risk factors for coronary heart disease with markedly improved serum lipid profile.[2] Almond skins are industrially removed from the nut by hot water blanching and represent 6-8% of the total almond weight. Currently they are under-utilised by-products of the almond processing industry and mainly used in cattle feed. However, almond skins are known to have a number of nutritional benefits, mainly based on the presence of polyphenols and the high dietary fibre content. A prebiotic is defined as "a nondigestible food ingredient which beneficially affects the host by stimulating the growth and/or activity of one or a limited number of bacteria in the colon and thus improving host health".[3] We have previously demonstrated that the bioaccessibility of nutrients and phytochemicals from almond seeds is improved by increased residence time in the gut and is regulated by almond cell walls.[4] Here we describe the modulation of human gut microbiota by blanched almond skins (BS) using a full model of gastrointestinal digestion which includes gastric and small intestinal environments and a colonic model of fermentation with representative human gut bacteria. Furthermore, the antimicrobial potential of polyphenols present in almond skins was evaluated against Gram-positive and Gram-negative bacteria, including food-borne pathogens.

2 METHOD AND RESULTS

2.1 *In vitro* digestion studies

For the *in vitro* digestion studies, BS were digested under gastric and duodenal conditions.[4] Briefly, for the gastric phase BS (1.5 g) were suspended in 12.4 ml acidic saline (150 mM NaCl, pH 2.5) in the presence of phosphatidylcholine (PC) vesicle suspension, pepsin and gastric lipase analogue at the concentration of 2.4 mM, 146 U/ml and 0.56 mg/ml, respectively. *In vitro* gastric digestion was performed for 2 h at pH 2.5. Following gastric digestion, the pH was immediately raised to 6.5 in order to simulate the duodenal conditions. Bile salts solution (sodium taurocholate and glycodeoxycholate), $CaCl_2$ and

BIS-TRIS buffer pH 6.5 were added to give final concentrations of 4 mM for each bile salt, 11.7 mM Ca Cl_2 and 0.73 mM BIS-TRIS. Duodenal digestions were initiated by addition of enzymes to give final concentrations of 5.9 BTEE U/ml for α-chymotrypsin, 104 BAEE U/ml trypsin, 3.2 μg/ml colipase and 54 U/ml pancreatic lipase. Digestions were performed in a shaking incubator (170 rpm) at 37°C for 1 h.

Table 1 *Chemical composition (expressed as g per 100 g) of blanched almond skins (BS) before and after in vitro gastric plus duodenal digestion. Values are means ± standard deviations. Other components include minerals and vitamins.*

Nutrient	Before *in vitro* gastric plus duodenal digestion	After *in vitro* gastric plus duodenal digestion
Total dietary fibre	45.1 ± 1.9	46.3 ± 1.5
Soluble dietary fibre	3.8 ± 0.5	4.5 ± 0.6
Lipid	24.2 ± 2.1	25.6 ± 2.0
Protein	12.8 ± 0.7	13.4 ± 1.0

Table 1 reports the gross composition of BS before and after gastrointestinal digestion, the latter subsequently used as substrates for the colonic model. As previously reported,[5] almond skins have high dietary fibre content which is not degraded in the upper GI tract. Only 3-4% was soluble (SDF) after digestion, the remaining being insoluble (ISF). Between 20 and 21% of protein and lipid were released from BS post *in vitro* gastric plus duodenal digestion, the gastric digestion step being responsible for the highest extent of lipolysis and proteolysis.

2.2 Colonic fermentation

Water-jacketed fermenter vessels (300-ml) filled with 135 ml of pre-sterilised basal growth medium (2 g/litre peptone water, 2 g/litre yeast extract, 0.1 g/litre NaCl, 0.04 g/litre K_2HPO_4, 0.04 g/litre KH_2PO_4, 0.01 g/litre $MgSO_4.7H_2O$, 0.01 g/litre $CaCl_2.6H_2O$, 2 g/litre $NaHCO_3$, 2 ml Tween 80, 0.02 g/litre hemin, 10 μl vitamin K_1, 0.5 g/litre cysteine HCl, 0.5 g/litre bile salts, pH 7.0) were inoculated with 15 ml of faecal slurry, prepared by homogenizing 10% (w/v) freshly voided faecal material of one healthy donor in 0.1 M PBS, pH 7.0. BS post digestion or fructo-oligosaccharides (FOS) was added to give a final concentration of 1% (w/v). Each vessel was magnetically stirred, the temperature set at 37°C, and pH automatically maintained at pH 6.8. Anaerobic conditions were maintained by sparging the vessels with oxygen-free nitrogen at 15 ml/min. Samples (5 ml) were removed at 0, 4, 8, and 24 h for bacterial enumeration and fatty acid analysis. Fermentations were run on three separate occasions.

Bacteria were counted using fluorescent *in situ* hybridization (FISH) as previously reported.[6] Differences between bacterial numbers at 0, 8 and 24 h of fermentation for each batch culture were checked for significance by paired *t*-test, assuming normal distribution, equal variances and considering both sides of distribution. The differences were considered significant when P was <0.05.

The relative changes in the number of different groups of bacteria after 8 and 24 h of incubation as a result of BS and FOS addition are shown in Figure 1 and Figure 2. An increase in the numbers of bifidobacteria and *Eubacterium rectale* was observed in response to the addition of BS and FOS at both 8 and 24 h incubation time points. The bacteroides and the clostridia population significantly decreased with both BS and FOS after 8 and 24 h fermentation. Addition of FOS produced and increase in the lactobacilli numbers after 8 and 24 h fermentation, whereas a slight decrease was observed after fermentation with BS.

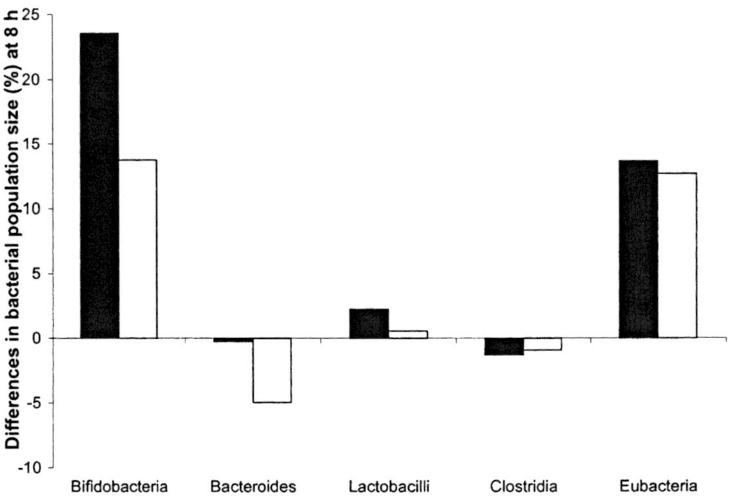

Figure 1 *Differences in the bacterial population size [black bars fructo-oligosaccharides (FOS); white bars blanched almond skins (BS)] compared to the total bacteria counted at 8 h [(selected bacterial numbers at 8 h/total bacteria counted at 8 h)-(selected bacterial numbers at 0 h/total bacteria counted at 0 h)].*

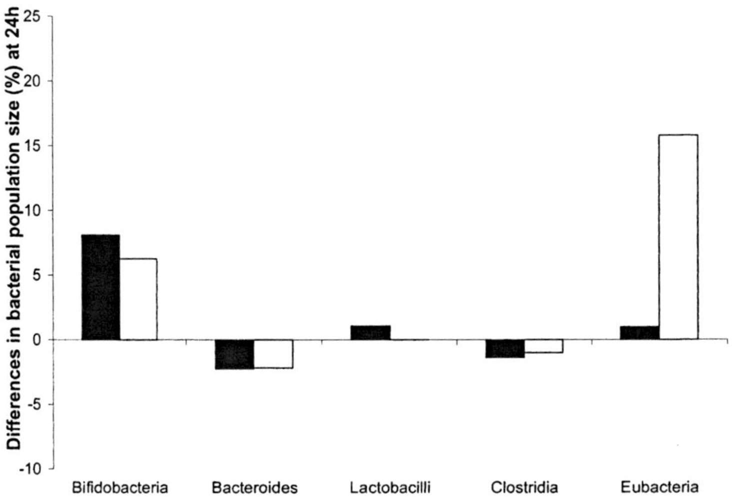

Figure 2 *Differences in the bacterial population size [black bars fructo-oligosaccharides (FOS); white bars blanched almond skins (BS)] compared to the total bacteria counted at 24 h [(selected bacterial numbers at 8 h/total bacteria counted at 8 h)-(selected bacterial numbers at 0 h/total bacteria counted at 0 h)].*

2.3 Antimicrobial potential

Table 2 *MIC values of a flavonoid-rich extract from BS against Gram-positive and Gram-negative bacteria. Values are expressed as µg/ml and represent the average of three independent determinations.*

Strain	MIC (µg/ml)
Enterococcus hirae	> 2000
Salmonella thyphi 0901	> 2000
Salmonella thyphimurium ATCC 14028	> 2000
Escherichia coli ATCC 25922	> 2000
Enterococcus durans	> 2000
Listeria monocytogenes ATCC 7644	500
Serratia marcescens	> 2000
Staphylococcus aureus ATCC 6538	500
Pseudomonas aeruginosa ATCC 27853	> 2000

The minimum inhibitory concentrations (MICs) of a flavonoids-rich extract derived from BS against Gram-positive and Gram-negative bacteria were determined by the method of broth microdilution, according to the guidelines established by NCCLS. [7] The main identified flavonoids in BS were protochatecuic acid, naringenin, catechin, epicatechin and isorhamnetin. BS was active against the food-borne pathogens *Listeria monocytogenes* ATCC 7644 and *Staphylococcus aureus* ATCC 6538 (Table 2). No antimicrobial potential was observed against the Gram-negative strains tested.

3 CONCLUSION

We have shown that addition of BS, a by-product of the almond processing industry, in batch fermentation cultures led to a significant increase in beneficial bacteria in the colon, such as bifidobacteria and eubacteria. Therefore BS shows potential for use as a functional food ingredient. Polyphenols present in almond skins are effective against Gram-positive food-borne pathogens and could be used as a possible source of natural antimicrobials.

References

1 http://www.almondboard.com
2 J. Sabate, E. Haddad, J.S. Tanzman, P. Jambazian and S. Rajaram, *Am. J. Clin. Nutr.*, 2003, **77**, 1379.
3 G.R. Gibson and M.B. Roberfroid, *J. Nutr.*, 1995, **125**, 1401.
4 G. Mandalari, R. M. Faulks, G.T. Rich, V. Lo Turco, D.R. Picout, R.B. Lo Curto, G. Bisignano, P. Dugo, G. Dugo, K.W. Waldron, P.R. Ellis and M.S.J. Wickham, *J. Agric. Food Chem.*, 2008, **56**, 3409.
5 G. Mandalari, A. Tomaino, T. Arcoraci, M. Martorana, V. Lo Turco, F. Cacciola, G.T. Rich, C. Bisignano, A. Saija, P. Dugo, K.L. Cross, M.L. Parker, K.W. Waldron and M.S.J. Wickham, *J. Food Comp. Anal.*, 2009 (in press).
6 G. Mandalari, C. Nueno-Palop, G. Bisignano, M.S.J. Wickham and A. Narbad, *Appl. Environ. Microbiol.*, 2008, **74**, 4264.
7 National Committee for Clinical Laboratory Standards. Performance Standards for Antimicrobial Susceptibility Testing: 17th International Supplement Clinical Laboratory Standards Institute, Wayne, PA, 2007, **M100**, S17.

APPLE (*MALUS DOMESTICA* BORKH. cv BRAMLEY'S SEEDLING) PEEL WASTE AS A VALUABLE SOURCE OF NATURAL PHENOLIC ANTIOXIDANTS

L. Massini, A.B. Martín Diana, C. Barry Ryan and D. Rico

Functional Ingredients Food Unit (FIFU), School of Food Science and Environmental Health, Dublin Institute of Technology (DIT), Dublin, Ireland.

1 INTRODUCTION

The traditional Irish apple market based on the cooking variety cv Bramley's Seedling has narrowed over the years due to changes in lifestyle. The peel waste was investigated as a potential source of natural phenolic antioxidants for food applications, in order to enhance the apple market through a recycling opportunity which may lower the cost impact of waste disposal. Depending on the variety, the peels of some apple cultivars have already been reported as a valuable functional source due to strong antioxidant capacity[1] and various inhibitory properties (i.e. anti-cancer, anti-allergic and anti-thrombotic)[2] which have been related to a higher phenolic content, in comparison with the peeled fruit.[3] In particular, the skin contains larger amounts of specific flavonoids such as flavonol glycosides, apart from phenolics found in the whole apple.[4] The antioxidant capacity of the fruit was also mainly attributed to its phenolic compounds, rather than to vitamin C.[5] With regard to the antioxidant capacity, there is still little literature data reporting about the value of the apple peels from cooking varieties,[6] particularly in the case of cv Bramley's Seedling, thus creating a research gap which is worth further investigation.

2 METHODS AND RESULTS

2.1 Quantification of Extractable Phenolic Compounds with Antioxidant Capacity in cv Bramley's Seedling Apple Peels

2.1.1 Contribution of the Peels to the Total Phenolic Content of the Fruit. Peel samples were prepared at lab scale from apples purchased from a local store and processed within a week of purchase. After removing the peels with a hand-peeler, the peeled fruits were crushed and juiced to produce a pomace which was used for comparison of total phenolic (TP) content with the peels. Both samples were oven-dried in a ventilated oven at 60°C (until a constant weight was obtained), then powdered. Polyphenols were extracted from apple flours by homogenising with aqueous ethanol 80% (v/v).

The ratio of total phenolics (i.e. assessed using Folin-Ciocalteu assay, see ref. 6 for assay method) of the peels/peeled fruit was 5.42 ± 0.35, on a dry weight (DW) basis, in agreement with data reported for other apple varieties.[7] In particular, the amount (%) of polyphenols in the peels was 81.94 ± 0.85%, compared to the TP content of the 'peel + pomace' samples. These findings were in agreement with previous data presented in literature, which has reported a higher phenolic content in the peels of different apple varieties (i.e. cv Rome Beauty, a cooking apple), compared to flesh or 'flesh + peel' samples.[8]

2.1.2 Effect of Preserving Treatments on the Recovery of Peel Polyphenols with Antioxidant Capacity. The reduction of the moisture content of fresh plant waste through a proper drying method is necessary to preserve it in view of recovery of phytochemicals for food applications.[9] The preserving effect of oven-drying on Bramley apple peel waste was evaluated, on a fresh weight (FW) basis, by comparing the amount of phenolics with antioxidant (AO) capacity recovered with aqueous ethanol from fresh and dried peels. The phenolic content was assessed as TP content and total flavonoid (TF) content (i.e. using Aluminium-chloride assay, see ref. 6 for assay method). The AO capacity was assessed as FRAP (i.e. Ferric Reducing Antioxidant Power)[10] and antiradical capacity against $DPPH^{\cdot}$,[11] using ascorbic acid as a reference antioxidant. The IC_{50} concentration of the peel extracts which enabled 50% inhibition against $DPPH^{\cdot}$ was estimated from inhibition curves obtained with an exponential regression model, $Y = Ymax \cdot (1 - e^{-kx})$. The antiradical power (ARP) was calculated as $ARP = (1/IC_{50})$.[12]

Dried peels showed similar ($P>0.05$) TP and FRAP values to fresh peels, but lower ($P<0.05$) TF content and ARP (Table 1). Soluble conjugated phenolics (particularly, cinnamic acids) may become detectable after flour preparation due to release from fresh tissues,[13] thus countervailing the loss of phenolic compounds during drying. As the Aluminium-chloride method used to assess total flavonoids is specific for this class of compounds, thus not accounting for phenolic acids, peel flours may show lower TF content compared to fresh samples. Different classes of phenolic compounds may also show different AO capacity, probably due to different antioxidant mechanisms.[14] Similar phenolic content of peel flours compared to fresh samples (on a FW basis) may also be due to the variability of the fruit batches investigated.[15]

Table 1 *Phenolic content and AO capacity of fresh and dried peels.*

Parameter	*Fresh peels*	*Dried peels*
TP content (mg GAE/100 g peels, FW)	634.99 ± 76.03^{a}	585.85 ± 34.72^{a}
TF content (mg CE/100 g peels, FW)	475.95 ± 47.18^{b}	403.12 ± 5.96^{a}
FRAP (mg AAE/g peels, FW)	9.88 ± 1.66^{a}	9.50 ± 0.42^{a}
ARP (mg peels/L, FW, in the final reaction mixture)$^{-1}$	234^{b}	155^{a}

GAE: gallic acid equivalents. CE: (+)-catechin equivalents. AAE: ascorbic acid equivalents. All values were presented as mean ± SD, n = 3, except for ARP (i.e. values were presented as fitted values without SD). Different letters indicate significant difference ($P<0.05$) between samples.

Among the possible strategies to reduce the enzymatic browning of a processed plant material, dipping in organic acids[16] or blanching have already been reported.[17] In order to increase the recovery yield of phenolic antioxidants from dried peels, different pre-oven treatments were applied before flour preparation and further ethanolic extraction, such as: dipping in ascorbic acid 0.5% (w/v) + NaCl 0.05% (w/v), for 3 minutes; water blanching for 10 seconds at 100°C; or steam blanching for 10 seconds, 30 seconds or 1 minute.

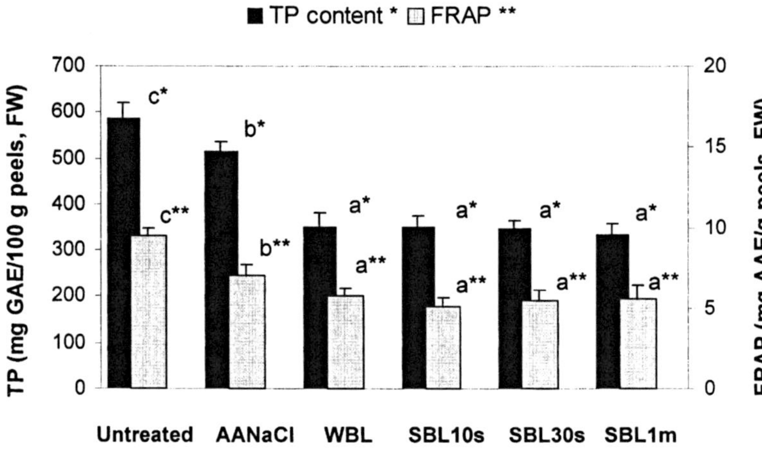

Figure 1 *Phenolic content and AO capacity of untreated and pre-oven treated peel flours. Pre-oven treatments: AANaCl (ascorbic acid + NaCl); WBL (water blanching); SBL1m (steam blanching, 1 min); SBL30s (steam blanching, 30 sec); SBL10s (steam blanching, 10 sec). The series of data for TP content and FRAP (TF and ARP values not shown) were distinguished with one or two asterisks, respectively. Different letters indicate significant difference (P<0.05, by analysis of variance) among treatments, for each series of data.*

Pre-oven treatments did not increase the amount of phenolics recovered from apple peels when compared to untreated samples. Pre-treated samples also showed lower antioxidant capacity in comparison with untreated samples (Figure 1). Among the pre-treatments, the effect of steam blanching or water blanching at the selected time conditions was similar (P>0.05), but less effective than dipping in ascorbic acid and sodium chloride. Apart from a possible loss of water soluble phytochemicals during blanching that has already been reported in processed plant material,[18] these findings may indicate that the selected pre-treatment conditions were unsuitable to the enzymatic profile of the cv Bramley's Seedling, with regard to its phenolic composition. Moreover, the applied oven-drying system with forced-air ventilation may enable a better deaeration, thus reducing the browning with a faster loss of moisture content. The effect may be less evident for apple peels which had been subjected to wet pre-treatments, before oven-drying.

The Pearson correlation coefficient between phenolic content and AO capacity of fresh and dried peels was statistically significant (P<0.05). TF was better correlated with DPPH, r = 0.93. TP showed higher correlation with FRAP, r = 0.92. These findings were in agreement with results already reported for other apple varieties (e.g. cv Golden Delicious, cv Ida Red, cv Northern Spy, and cv Red Delicious).[19]

2.1.3 *Use of a Hydrolysis Method for Releasing Insoluble Bound Peel Phenolics.* After the ethanolic extraction of soluble polyphenols, the presence of phenolics which may be bound to the cell wall material, thus insoluble in the organic solvent, was investigated in the residual peel flour (i.e. referred to as spent peel flour) by applying a basic hydrolysis for 4 hours, under nitrogen, with NaOH 4M. After neutralisation, it has been observed that the AO capacity (i.e. assessed as FRAP) of the hydrolysates was significantly ($P<0.05$) correlated with TP and TF content ($r = 0.95$ and $r = 0.97$, respectively). These results may suggest the presence of saponifiable phenolics in the spent peel flour. By comparing the phenolic content of the hydrolysates with crude peel extracts, it has been shown that the yield of soluble polyphenols accounted for more than 70% of total phenolics in the peels (i.e. assessed as soluble and insoluble compounds), on a FW basis. Crude peel extracts also showed higher AO capacity ($P<0.05$) compared to the hydrolysates. As already reported in literature for other apple varieties,[20] most of the polyphenols in cv Bramley's Seedling peels were soluble, thus not bound to the cell wall material. The hydrolysis procedure could be omitted after the liquid extraction, making the recovery of phenolic antioxidants from peel waste more feasible for industrial food applications.

2.2 Recovery of Phenolic Fractions and Identification of Target Compounds with High Antioxidant Capacity in cv Bramley's Seedling Apple Peels

Purified fractions of neutral and acidic polyphenols were obtained from crude peel extracts by using solvents of different polarity and a pH shifting method, before partitioning with ethyl acetate (Figure 2).[21]

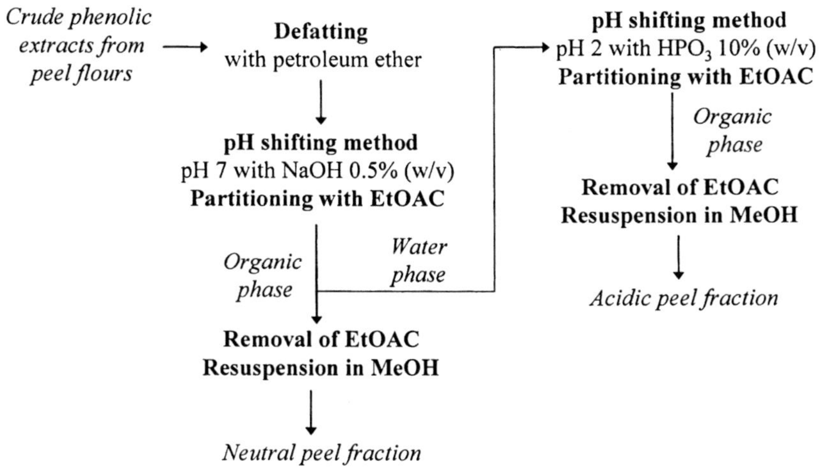

Figure 2 *Fractionation scheme for obtaining neutral and acidic phenolics from crude peel extracts. EtOAC: ethyl acetate. MeOH: methanol.*

Fractions were analysed with colorimetric methods for phenolic content, including total flavan-3-ols content and relative degree of polymerisation (rDP) of procyanidins (i.e. assessed with Vanillin assay and Acid Butanol-Vanillin assay, respectively),[22] and FRAP and DPPH˙ antiradical capacity. Fractions were also tested as *in vitro* radical scavengers against hydroxyl radicals (i.e. with Deoxyribose assay).[23] Fractions were then characterised by RP-HPLC analysis, using a Nucleosil C18 column and an acetic acid/acetonitrile gradient (with modifications, see ref. 21 for method), in order to identify target classes of phenolic compounds with high AO capacity.

The use of a pH shifting method enabled the recovery of two phenolic fractions with different phenolic content and AO capacity (Figure 3). In particular, the neutral fraction showed higher phenolic content and higher AO capacity ($P<0.05$) compared to the acidic fraction. The higher content of flavan-3-ols (data not shown) in the neutral fraction was characterised by a lower rDP (i.e. 1.51 ± 0.08) which may indicate the presence of dimers or monomers. The acidic fraction showed a lower content of flavan-3-ols but with a higher rDP (i.e. 2.67 ± 0.33), possibly indicating the presence of oligomeric procyanidins. Both peel fractions were active as *in vitro* radical scavengers against hydroxyl radicals. The neutral fraction showed stronger capacity over the acidic one, at different concentrations (data not shown).

Comparison with phenolic standard retention times and literature data was used in the RP-HPLC analysis of the peel fractions.[24] The neutral fraction was enriched with flavonoids, such as flavonol glycosides, dihydrocalchones and flavan-3-ols. In particular, (+)-catechin and phloridzin were identified, while procyanidin B2, (-)-epicatechin and quercetin glycosides were suggested in view of further analysis. The acidic fraction was characterised by the presence of hydroxycinnamic acids. Chlorogenic acid was identified, while another phenolic acid, not yet identified, was separated. Relatively smaller peaks with UV spectral characteristics similar to those of flavan-3-ols were shown to be present in the acidic fraction, as suggested by colorimetric methods. Further separation is needed before drawing final conclusions.

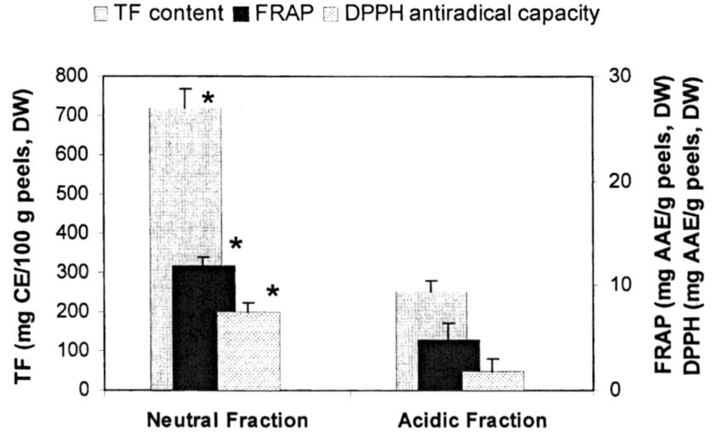

Figure 3 *Phenolic content and AO capacity of neutral and acidic peel fractions. The asterisks indicate significant difference (P<0.05) between fractions.*

3 CONCLUSION

The antioxidant capacity of fresh and dried peels from cv Bramley's Seedling apple was significantly correlated with the phenolic content. In view of industrial food applications, a feasible system such as oven-drying with air ventilation was used to preserve the antioxidant potential of the peel waste, by applying a food-grade liquid extraction with aqueous ethanol that has recovered most of the polyphenols, which were soluble, not bound to the cell wall material. The use of a simple fractionation procedure with pH shifting method and ethyl acetate partitioning allowed the recovery of a neutral phenolic fraction with higher flavonoid content and stronger AO capacity over the acidic one, thus being promising for further food applications. Both fractions showed an *in vitro* hydroxyl radical scavenging capacity, which may be worth further investigation as a possible bioactivity for functional purpose.

References

1 F. Chinnici, A. Bendini, A. Gaiani and C.S. Riponi, *J. Agric. Food Chem.*, 2004, **52**, 4684.
2 P.C.H. Hollman and M.B. Katan, *Food Chem. Toxicol.*, 1999, **37**, 937.
3 M.V. Eberhardt, C.Y. Lee and R.H. Liu, *Nature*, 2000, **6789**, 903.
4 S. Burda, W. Oleszek and C.Y. Lee, *J. Agric. Food Chem.*, 1990, **38**, 945.
5 K.W. Lee, Y.J. Kim, D.-O. Kim, H.J. Lee and C.Y. Lee, *J. Agric. Food Chem.*, 2003, **51**, 6516.
6 K.L. Wolfe and R.H. Liu, *J. Agric. Food Chem.*, 2003, **51**, 1676.
7 B. Łata, *J. Agric. Food Chem.*, 2007, **55**, 663.
8 K. Wolfe, X. Wu and R.H. Liu, *J. Agric. Food Chem.*, 2003, **51**, 609.
9 A. Schieber, F.C. Stintzing and R. Carle, *Trends Food Sci. Tech.*, 2001, **12**, 401.
10 Z. Maksimović, Đ. Malenčić and N. Kovačevića, *Bioresource Technol.*, 2005, **96**, 873.
11 S. Kondo, K. Tsuda, N. Muto and J.-E. Ueda, *Sci. Hortic.*, 2002, **96**, 177.
12 W. Brand-Williams, M.E. Cuvelier and C. Berset, *Lebensm. Wiss. Technol.*, 1995, **28**, 25.
13 J.A. Vinson, X. Su, L. Zubik and P. Bose, *J. Agric. Food Chem.*, 2001, **49**, 5315.
14 R.L. Prior, X. Wu and K. Schaich, *J. Agric. Food Chem.*, 2005, **53**, 4290.
15 W. Peschel, F. Sanchéz-Rabaneda, W. Diekmann, A. Plescher, I. Gartzía, D. Jiménez, R. Lamuela-Raventós, S. Buxaderas and C. Codina, *Food Chem.*, 2006, **97**, 137.
16 S. Lu, Y. Luo, E. Turner and H. Feng, *Food Chem.*, 2007, **104**, 824.
17 W. Kalt, *J. Food Sci.*, 2005, **70**, 11.
18 M.I. Gil, F. Ferreres and F.A. Tomás-Barberán, *J. Agric. Food Chem.*, 1999, **47**, 2213.
19 R. Tsao, R. Yang, S. Xie, E. Sockovie and S. Khanizadeh, *J. Agric. Food Chem.*, 2005, **53**, 4989.
20 J. Sun, Y.-F. Chu, X. Wu and R.H. Liu, *J. Agric. Food Chem.*, 2002, **50**, 7449.
21 A. Schieber, P. Keller and R. Carle, *J. Chromatogr. A*, 2001, **910**, 265.
22 A.E. Hagerman and L.G. Butler, *Meth. Enzymol.*, 1994, **234**, 429.
23 B. Halliwell, J.M.C. Gutteridge and O.I. Aruoma, *Anal. Biochem.*, 1987, **165**, 215.
24 A. Escarpa and M.C. González, *J. Chromatogr. A*, 1999, **830**, 301.

OPTIMIZATION OF CULTIVATION CONDITIONS FOR THE PRODUCTION OF BACTERIAL PHYTASE FROM *ENTEROBACTER SAKAZAKII* ASUIA279 NEWLY ISOLATED FROM MALAYSIAN MAIZE ROOT

Anis Shobirin Meor Hussin,[1] Abd-ElAziem Farouk,[2] Hamzah Salleh,[3] Abd Manaf Ali,[4] and Aini Ideris[5]

[1]Faculty of Food Science and Technology, University Putra Malaysia, 43400, UPM, Serdang, Selangor.
[2] Faculty of Science, University of Taif, Taif, Al-Hawiya 888, Saudi Arabia.
[3]Kulliyyah of Engineering, International Islamic University, Faculty of Engineering, Jalan Gombak, 53100 Kuala Lumpur, Malaysia.
[4]Faculty of Agricultural and Biotechnology, Universiti Darul Iman Malaysia.
[5]Faculty of Veterinary Medicine, Universiti Putra Malaysia Putra Malaysia.

1 INTRODUCTION

Approximately 70% of total phosphorus in cereal grains and legume seeds exists as phytate. The phytate is poorly utilized by monogastric animals and acts as an anti-nutrient due to its chelating of various metals and binding of protein. This diminishes the bioavailability of proteins and nutritionally important minerals.

The phytate degrading enzymes or phytases convert phytate to inositol and phosphoric acid, making phosphorus available for bioabsorption.[1] Supplementation of microbial phytase to animal diets alters the phytate complexes and also increases the bioavailability of proteins and essential minerals, providing growth performance equivalent or better than those with phosphate supplementation, and also reduces the amount of phosphorus in animal manure.[2]

Partial optimization of the cultivation conditions for the productivity is becoming one of the major goals in biotechnology today. The productivity of any cultivation is affected by process parameters and media composition.[3] Preliminary experiments in our laboratory indicated that rice bran composition and physical parameter of cultivation conditions play a significant role in phytase production. Therefore an investigation was performed to statistically optimize the cultivation conditions for the production of the highly potential extracellular phytase from *Enterobacter sakazakii* ASUIA279. Response surface methodology (RSM) is now being routinely used for optimization studies in several biotechnological and industrial processes[4–6] RSM is used to explain the interaction among the possible influencing parameters with limited numbers of parameters.

In this paper, we determine the optimum composition of rice bran supplement in the media, initial pH of the media, incubation temperature, agitation and aeration during cultivation for phytase production by using full factorial Face Centred Central Composite Design in RSM.

2 MATERIALS AND METHODS

2.1 Microorganism and Culture Conditions

Enterobacter sakazakii ASUIA279 was isolated from the endophyte of Malaysian Maize root and identified using biochemical and phylogenetic analysis.[7] The strain was maintained on agar slopes at 4 °C. LB broth (Merck) with initial pH of 7.0 in universal bottle used as the standard inoculum medium and kept in incubating shaker at 37 °C and 200 rpm for 18 h.

2.2 Chemicals

Phytic acid, dodecasodium salt, was purchased from Sigma Chemical Co. (St. Louis, Mo.). All other chemicals and media were of analytical grade.

2.3 Experimental Design

After preliminary optimizing the values of various nutritional and physical factors in individually, the five most important factors, namely rice bran concentration (X_1), incubation temperature (X_2), initial pH (X_3), agitation speed (X_4) and aeration volume (X_5) were the most influential factors for phytate-degrading enzyme production by *Enterobacter sakazakii* ASUIA279 under batch fermentation conditions. Taking the above factors and constraint of time into consideration, a response surface methodology using a two-step experimental design was adopted for improving total phytate-degrading enzyme secretion from *Enterobacter sakazakii* ASUIA279. The experiments involved the phytate-degrading enzyme production for 20 run in Erlenmeyer flask (see table 2) and then scaling up in a bioreactor for 13 run (see table 3).

In step-one, 10 ml of inoculums (10 % v/v) was transferred into 250 ml Erlenmeyer flasks containing 100 ml production medium for optimization of the rice bran concentration, the initial pH of media and the incubating temperature. The flasks were then placed in an orbital shaker at 200 rpm. The experiments were done in triplicate. While in step-two, 2 L bioreactor (Biostat® B-DCU- B. Braun Biotech International, Sartorius Group) containing 1 L production medium was used for optimization of the agitation and the aeration. The production medium, the initial pH of the media and the incubation temperature was the optimized conditions derived from step-one. The samples of each one were analyzed three times. Samples of 1.5 ml were withdrawn from both steps after 5 days of incubation.

Full-factorial Central Composite Experimental Design with centred face was used to optimize the cultivation conditions for phytate-degrading enzyme production by *Enterobacter sakazakii* ASUIA279. The experiments were designed by using the Minitab 14 Statistical Software. A 2^3 – factorial experimental design with four axial points (with $\alpha = 1$ and six replicates at the centre point ($n_0 = 6$) leading to a total number of twenty experiments was employed for the optimization of rice bran composition (X_1), initial pH (X_2) of media and incubation temperature (X_3) in Erlenmeyer flasks and a 2^2 – factorial experimental design leading to thirteen trials for the optimization of agitation (X_4) and aeration (X_5) in a 2L bioreactor.

The range of the variable choosen is based on other studies.[8] The average of the maximum phytate-degrading enzyme activity of the duplicate values obtained was taken as dependent variable or response Yi (U). Regression analysis and isoresponse contour plots were performed on the data obtained using MINITAB 14. Microsoft Office Excel 2003 is used to calculate the predicted responses.

The final validation of the statistical regression model for prediction of phytate-degrading enzyme production from *Enterobacter sakazakii* ASUIA279 was carried out in both steps. Samples were withdrawn for phytase assay after 5 days of cultivation.

3 RESULTS

From the preliminary study, the factors that have the major influence for *Enterobacter sakazakii* ASUIA279 to produce phytate-degrading enzyme were the composition of rice bran, incubation temperature, initial pH, agitation and aeration. Due to apparatus and time constraints, the experiments were split into two steps where the first three factors were optimized in Erlenmeyer flask. With the optimized cultivation conditions derived from the first step, the agitation and aeration were optimized in a 2 L bioreactor

The response was taken at the maximum phytate-degrading enzyme activity, which was observed after 5 days of fermentation. Table 1 and 2 summarize the phytate-degrading enzyme activity (response) of the experiments for both steps. The coefficients of the regression equation were calculated using MINITAB 14 and the following regression equations were obtained.

Step-one;

$$Y = -44.4 + 1.15X_1 + 6.96X_2 - 0.165X_3 - 0.0212X_1^2 - 0.759X_2^2 - 0.0439X_3^2 + 0.0991X_1X_2 + 0.0299X_1X_3 + 0.226X_2X_3 - 0.00595X_1X_2X_3 \quad (1)$$

$R^2 = 97.4\%$

Step-two;

$$Y = -7.03 + 0.102X_4 - 3.93X_5 - 0.000156X_4^2 + 0.586X_5^2 - 0.00172X_4X_5 \quad (2)$$

$R^2 = 98.3\%$

Where Y= the response or phytate-degrading enzyme activity

X_1= temperature (ºC)

X_2= pH

X_3= rice bran (% w/v)

X_4= agitation (rpm)

X_5= aeration (vvm)

The significance of each coefficient was determined by T values and P values which are listed in table 3 and 4. From t-table, alpha value of 0.05 and degree of freedom of 19, the critical value of t is 1.73 (step-one) while for degree of freedom of 12, the critical value of t is 1.78 (step-two). Therefore, factors with t values greater than t critical are significant to the model. Based on the t-test results, it is observed from table 6 that the coefficient for temperature (X_1) and initial pH (X_2) are the most significant factors for step-one. Furthermore, the quadratic interactions X_1^2, X_2^2, X_3^2 and the factor of interactions X_1X_2 (temp, pH), X_2X_3 (pH, % rice bran), $X_1X_2X_3$ (temp, pH, % rice bran) were also found to be very significant. Table 5 shows that the coefficient for both agitation speed (X_4) and aeration (X_5) as well the quadratic interaction X_4^2 and X_5^2 are significant factors for step-two. These significant factors are very important as they indicates the limiting factors in which the smallest variations in their values can alter the phytate-degrading enzyme production.

Table 1 *Experimental design of three independent variables in step-one showing experimental and predicted responses.*

Experiment run no.	Temperature (°C)	pH	% Rice Bran	Phytase Activity (U/ml)	
1	37	7	10	12.009	11.212
2	45	5	5	6.360	6.704
3	45	5	15	7.227	6.096
4	37	7	10	12.408	11.212
5	29	5	5	5.030	4.795
6	37	7	10	12.058	11.212
7	29	7	10	8.708	8.325
8	37	7	5	9.393	10.609
9	37	7	15	10.506	11.457
10	45	7	10	10.673	10.695
11	45	9	5	8.696	7.843
12	37	9	10	8.118	7.935
13	29	5	15	5.883	5.794
14	37	7	10	11.209	11.212
15	37	5	10	7.905	7.729
16	37	7	10	11.839	11.212
17	37	7	10	12.197	11.212
18	29	9	5	2.931	2.762
19	29	9	15	5.903	5.710
20	45	9	15	7.871	7.897

Table 2 *Experimental design of two independent variables in step-two showing experimental and predicted responses.*

Experiment run no.	Agitation (rpm)	Aeration (vvm)	Phytase Activity (U/ml)	
			Experimental	Predicted
1	300	0	10.237	9.450
2	300	1	5.583	5.596
3	400	2	2.108	1.994
4	300	1	5.891	5.596
5	300	1	5.460	5.596
6	200	0	6.835	7.25
7	200	2	1.229	0.714
8	400	0	8.402	8.53
9	400	1	5.011	4.676
10	300	1	5.664	5.596
11	300	2	2.515	2.914
12	200	1	3.448	3.396
13	300	1	5.545	5.596

Table 3 *The least-squares fit and parameter estimates (significance of regression coefficient) for step-one.*

Predictor	Coefficient	SE Coefficient	T	P
Constant	-44.44	11.140	-3.99	0.003
X_1	1.148	0.505	2.27*	0.049
X_2	6.960	1.815	3.84*	0.004
X_3	-0.165	0.843	-0.20	0.849
X_1^2	-0.021	0.006	-3.49*	0.007
X_2^2	-0.759	0.097	-7.82*	0.000
X_3^2	-0.044	0.016	-2.83*	0.020
X_1X_2	0.099	0.032	3.11*	0.012
X_1X_3	0.030	0.021	1.44	0.183
X_2X_3	0.226	0.108	2.10*	0.066
$X_1X_2X_3$	-0.006	0.003	-2.09*	0.066

* indicates significance to the model

Table 4 *The least-squares fit and parameter estimates (significance of regression coefficient) for step-two.*

Predictor	Coefficient	SE Coefficient	T	P
Constant	-7.030	2.302	-3.05	0.018
X_4	0.102	0.016	6.53*	0.000
X_5	-3.927	0.837	-4.69*	0.002
X_4^2	-0.0001	0.000	-6.09*	0.000
X_5^2	0.586	0.256	2.29*	0.056
X_4X_5	-0.002	0.002	-0.81	0.446

* indicates significance to the model

The coefficients of the regression equation were re-analysed and calculated using MINITAB 14 and the following regression equations were obtained.

Step-one,

$$Y = -61.5 + 1.84X_1 + 9.35X_2 - 0.0266X_1^2 - 0.845X_2^2 - 0.00718X_3^2 + 0.0596X_1X_2 + 0.107X_2X_3 - 0.00201X_1X_2X_3 \quad (3)$$

$R^2 = 94.7\%$

Step-two,

$$Y = -6.51 + 0.100X_4 - 4.44X_5 - 0.000156X_4^2 + 0.586X_5^2 \quad (4)$$

$R^2 = 98.1\%$

Where Y= the response or phytate-degrading enzyme activity

X_1= temperature (°C)

X_2= pH

X_3= rice bran (% w/v)

X_4= agitation (rpm)

X_5= aeration (vvm)

The regression equation and determination coefficient R^2 were evaluated to test the fit of the model. The model presented a high determination coefficient of 90.8 % for step-one and 98.1 % in step-two of the variability in the response.

The predicted responses were calculated using the re-analysed regression equation (see table 1 and table 2). The high value of R^2 indicating the high significance of the model. It revealed a good agreement between the experimental and predicted values of the responses. From the predicted responses, the optimum operating conditions for the enzyme production were at temperature of 37 °C, initial pH of 7, rice bran of 15%, 300 rpm of agitation speed and 0 vvm of aeration.

Figure 1 shows the contour plot of the relative effects of two most significant factors when all the factors were kept at their central levels. The plots depicted interaction of incubation temperature and initial pH for phytase production in step-one. The contour plot clearly shows an ellipse where the optimum conditions can be easily determined. These indicate that there are major interactions among the independent variable corresponding to the response surface. In the investigated area, it shows that the optimum level of cultivation conditions that produced highest phytate-degrading enzyme were at the incubation temperature of 39.7 °C, initial pH media of 7.1 and rice bran at 13.6%.

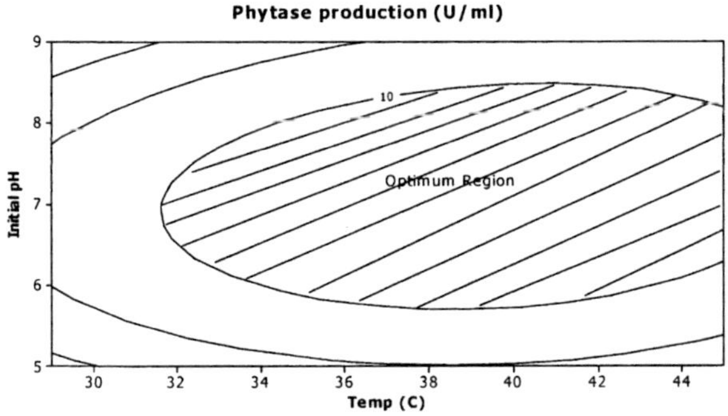

Figure 1 *Contour plot of phytate-degrading enzyme production from Enterobacter sakazakii ASUIA279 showing interaction between incubation temperature and initial pH*

The validation of model with experimental data as well the predicted values from response surface model under the optimized cultivation conditions (included only the most significant coefficient) are presented in table 8. Phytate-degrading enzyme productions under the optimized cultivation conditions for the validation model for both steps were within the range of the predicted value from the response surface model. The optimization of these parameters using RSM has increased the production of phytate-degrading enzyme by *Enterobacter sakazakii* ASUIA279. There are no known reports on statistically optimization of temperature, initial pH, rice bran concentration, agitation and aeration for production of phytase. Quan *et al.*[9] and Gautam *et al.*[10] had done an individual optimization (keeping other parameters as constant) on the effect of temperature and pH besides other parameters for phytase production. The effect of solid substrates (canola meal, corn meal, soy meal and wheat meal) has been optimized using Plackett- Baurman Design by Bogar *et al.*[11]

Table 5 *Experimental (validation studies) and predicted results (from Response Surface Model) of phytate-degrading enzyme activity (U/ml) under optimized cultivation condition.*

Phytase Activity in Erlenmeyer Flask (U/ml)		Phytase Activity in Bioreactor (U/ml)	
Experimental	Predicted	Experimental	Predicted
11.14 ± 0.12	11.511	9.56 ± 0.09	9.516

4 DISCUSSION

The conventional methods for multifactor experimental design are time-consuming and incapable of obtaining the true optimum, due especially to the interactions among the factors.[12] In order to design a suitable bioprocess for optimize phytate-degrading enzyme production from *Enterobacter sakazakii* ASUIA279, the statistical design approach using response surface methodology with face centred central composite design (FCCCD) was used to study the interactive effects of various nutritional and physical factors on phytate-degrading enzyme production. According to Vohra and Satyanarayana,[8] the source and optimal concentration of carbon are important factors for the production of phytate-degrading enzyme and the most important physical parameters having profound influence on growth of organisms and production of metabolites from them are pH, temperature, agitation, dissolved oxygen, and pressure.

Student's t-test was used to determine the significant coefficient of the experiment run. If the T value observed from the regression coefficient data was higher or equal to critical t obtained from t-table, indicated that the coefficient was significant to the model. For step-one, at $\alpha = 0.05$, degree of freedom = 19, and the critical value of t is 1.73, its shows that the most significant factor that effect the phytate-degrading enzymes were the incubation temperature and the initial pH. While for step-two, at $\alpha = 0.05$, degree of freedom = 12, and the critical value of t is 1.78, both agitation speed and aeration were significant factors that effect the production. The ANOVA of quadratic regression model demonstrates that the model is highly significant, as is evident from the Fishers F-test with very low probability value ($P > F = 0.0001$).

The larger the magnitude of the T value and the smaller the P value indicate the high significance of the corresponding coefficient.[13] It can be seen in step-one that the variable with the most effect was the quadratic effect of initial pH (X_2^2, T = 7.82, P = 0.000) followed by linear effect of pH (X_2, T = 3.84, P = 0.004) and than quadratic effect of temperature (X_1^2, T = 3.49, P = 0.007). The highest probability value of coefficient is the linear effect of rice bran percentage (0.838), indicating that only 16 % of the model is affected by this variable. Since linear and quadratic effect of initial pH is the most significant, meaning that it can act as limiting factor and little variation will alter the product formation rate. In the bioreactor, it shows that both linear effects and quadratic effects of agitation and aeration are highly significant (> 99 %).

The values of determination coefficient R^2 (step-one) = 90.8%, R^2 (step-two) = 98.1%and the adjusted determination coefficient R^2 (step-one) = 87.6%, adjusted determination coefficient R^2 (step-two) = 97.2% of the significant factors are very high, indicating a high significance of the model[14,15] which is more suited for comparing models with different numbers of independent variables.[16] This reveals a good agreement between the experimental and predicted values of phytate-degrading enzyme production.

The contour plots are graphical representations of the regression equation. The main goal of response surface is to hunt efficiently for the optimum values of the variables such that the response is maximized. Each contour curve represents an infinitive number of combination of two test variables with the other one maintained at their respective zero level. The maximum predicted value is indicated by the surface confined in the smallest ellipse in the contour diagram is the optimal region. Elliptical contours are obtained when there is a perfect interaction between the independent variables.[17]

From several studies[3,18] the optimum temperature for the production of phytate-degrading enzymes from most of the micro-organisms lies in the range 25 to 37 °C. Using the regression equation, the predicted most optimum level of cultivation conditions for higher phytate-degrading enzyme production was found to be at the incubation temperature of 39.7 °C, pH 7.1 and 13.6 % of rice bran concentration with 11.511 U/ml of enzyme yield. And when it was scale up in bioreactor for the optimization of agitation speed and aeration, the predicted optimum level of both factors were at 320 rpm and at 0 vvm it was 9.516 U/ml. A study done by Segueilha *et al.*[19], phytate-degrading enzyme production by *Schwannimyces castellii* was carried out continuously in a fermentor aerated at 1 vvm and agitation kept at 600 rpm. Phytate-degrading enzyme was produced maximally by *Bacillus* sp. DS11 by shaking at 230 rpm.[20] And, phytate-degrading enzyme produced by *Aspergillus niger, Aspergillus ficuum* NRRL 3135 and *Aspergillus terreus* was carried out by shaking at 270 rpm.[21] This indicates that optimum aeration and agitation are important to keep the cultivation conditions in a capable state.

The response surface model is a valuable tool for predicting and optimizing the cultivation conditions to maximize the phytate-degrading enzyme production. The results show that phytate-degrading enzyme production by *Enterobacter sakazakii* ASUIA279 can be improved by controlling various factors simultaneously. The validation model for both steps was in close agreement with the predicted value from the response surface model for phytate-degrading enzyme productions under the optimized cultivation conditions. The production of phytate-degrading enzyme was increased about 6-fold (compared to during screening) when cultivated at optimized condition in Erlenmeyer flask. However, the phytate-degrading enzyme production was decreased to 5-fold when the volume was scale-up in bioreactor. This could be due to the different surroundings in term of temperature, and volume between the two apparatus.

References

1 G.C.J. Irving and D.J. Cosgrove, *Austr. J. Biol. Sci.,* 1971, **24**, 547.

2 R.J. Wodzinski and A.M.J. Ullah Phytase. *Adv. Appl. Microbiol.* 1996, **42**, 263.

3 K.M. Park and K.F. Reardon, *Biotechnol. Lett.,*1996, **18**, 737.

4 Q.K, Beg, R.K. Saxena and R. Gupta, *Biotechnol. Bioeng.,* 2002, **78**, 289.

5 J. De Cornink, S. Bouquelet, V. Dumortier, F. Duyme and V.I. Denantes, *J. Ind. Microbiol. Biotechnol.* 2000, **24**, 285.

6 S. Puri, Q.K. Beg and R. Gupta, *Current Microbiology,* 2002, **44**, 286.

7 M.H Anis Shobirin, *PhD Thesis Dissertation*. IIUM, 2008.

8 A. Vohra and T. Satyanarayana, *Crit. Rev. Biotechnol.* 2003, **23**, 29.

9 C. Quan, L. Zhang, Y. Wang, and Y. Ohta, Journal *Biosci. Bioeng.,* 2001, **92**, 154.

10 P. Gautam, A. Sabu, A. Pandey, G. Szakacs and C. Soccol, *Bioresour. Technol.,* 2002, **83**, 229.

11 R. Bogar, G. Szakacs, J.C. Linden, A. Pandey and R.P. Tengerdy, *J. Ind. Microbiol. Biotechnol.,* 2003, **30**, 183.

12 T. Nagashima, T. Tange and H. Anazawa, *Appl. Environ. Microbiol.,* 1999, **65**, 4682.

13 R.S. Karthikeyan, S.K. Rakshit and A. Baradarajan, *Bioprocess Eng.* 1996, **15**, 247.

14 S. Akhnazarova and V. Kefarov, Experiment optimization in chemistry and chemical engineering. Moscow: Mir Publisher, 1982.

15 A.I. Khuri and J.A. Cornell, Response surfaces: design and analysis. New York: John Wiley and Sons, 1987 p 291.

16 Q.K. Beg, V. Sahai and R. Gupta, *Proceeding Biochemistry* 2003, **39**, 203.

17 R.V. Muralidhar, R.R. Chirumamila, R. Marchant and P. Nigam, *Biochem. Eng.,* 2001, **9**, 17.

18 R.M. Berka, M.W. Rey, K.M. Brown, T. Byun and A.V. Klotz, *Appl. Environ. Microbiol.,* 1998, **64**, 4423.

19 L. Segueilha, C. Lambrechts, H. Boze, G. Moulin and P. Galzy, *Fermentation Bioengineering* 1992, **74**, 7.
20 Y.O. Kim, H.K. Kim, K.S. Bae, J.H. Yu and T.K. Oh, *FEMS Microbiol. Lett.*, 1998, **162**, 185.
21 T.R. Shieh and J.H Ware, *Appl. Microbiol.*, 1968, **16**, 1348.

UTILIZATION OF PUMPKIN FLOUR IN EXPANDED BREAKFAST CEREALS

M.N. Norfezah,[1] C.S. Brennan[2] and A. Hardacre[1]

[1] Institute of Food, Nutrition and Human Health, College of Sciences, Massey University, Private Bag 11222,Palmerston North 4442 New Zealand
[2] Dept of Food and Tourism Management, Hollings Faculty, Manchester Metropolitan University, Old Hall Lane, Manchester, United Kingdom

1 INTRODUCTION

Pumpkin is a fruit that has a large range of uses as a valuable food source for humans and animals. The flesh is eaten freshly cooked as a vegetable or used as a flavour ingredient in pies, breads and soups. When dried as a powder or flakes it is used in foods ranging from sauces to jams. Moreover, it is a cheap and readily available source of vitamin A.[1] The versatility and profitability of the pumpkin crop could be enhanced by diversifying crop use and utilizing waste streams. Experience has shown that extrusion technology can be used to make tasty, nutritious snacks or breakfast cereals from pumpkin flour when blended with corn-grits. Therefore, to encourage greater consumption of pumpkin we consider that it may be possible to develop extruded breakfast cereals that incorporate nutritionally useful quantities of pumpkin flour. The objective of this study is to determine the use of pumpkin fractions as an ingredient for the manufacture of extrusion cooked expanded foods that could be suitable in breakfast cereals.

2 METHODS AND RESULTS

2.1 Methods

2.1.1 Raw Material. Matured crown and buttercup pumpkin (*Cucurbita maxima*) were purchased from the local supermarket.

2.1.2 Sample Preparation. Pumpkins were diced and separated into four fractions: peel, flesh, pulp and seed. Then all fractions were soaked in 0.1% citric acid for 30 minutes and drained. All four fractions of pumpkin (peel, flesh, pulp and seed) were transferred to wire-net trays and dried in the hot air drier for 24 hours at 65°C and 40% humidity. The dried fractions were ground into flour using a grinder and vacuum packed for further use.

2.1.3 Extrusion Process. Extrusion expanded breakfast cereals were made from peel, flesh and seed fractions of crown pumpkin flour were incorporated with 50%, 30% or 10%

w/w with 220 spec corn grit supplied by Corson Grain Ltd, (Gisborne, New Zealand). The blends were extruded using a Clextral BC21 twin-screw. Basic recipes and extrusion conditions were used as described in previous research.[2]

2.1.4 Chemical Composition. Moisture content and proximate analysis were determined according to the AOAC method.[3] All measurements were expressed on a wet matter basis and reported as a mean (± S.D.) of triplicate analysis.

2.1.5 Physical Characteristics. Expansion was determined by the ratio of radial diameter of the products to the die diameter and product density by the poppy seed displacement method.

2.1.6 Statistical Analysis. Statistical analysis were performed using SAS 9.1 (SAS Institute Inc, Cary, NC, USA) to calculate means ± standard deviation (S.D.) of values measured for each sample. All significant differences are reported at a significance level of $p < 0.01$.

2.2 Results

2.2.1 Moisture Content and Proximate Composition : The proximate composition and moisture content of pumpkin and pumpkin flour are presented in Table 1. Significant differences ($P<0.01$) were found in a number of areas in terms of type of pumpkin, drying method and fractions. For both varieties, drying methods showed a significant effect ($P<0.01$) on the moisture, ash, protein, fat and carbohydrate content.

Table 1 *Moisture content and proximate composition of pumpkin and pumpkin flour.*

Pumpkin	Processing	Fraction	Moisture	Ash	Protein	Fat	Carbohydrate
Crown	Raw	Peel	89.52^{b}	9.83^{a}	0.32^{m}	0.03^{f}	0.30^{j}
		Flesh	84.34^{a}	8.20^{b}	0.29^{m}	0.28^{c}	6.89^{k}
		Seed	47.85^{e}	3.45^{e}	2.85^{j}	0.09^{f}	45.76^{i}
	Cabinet Drying	Peel	10.31^{i}	6.50^{d}	1.96^{k}	0.27^{c}	80.95^{c}
		Flesh	14.43^{h}	5.62^{e}	1.30^{l}	0.04	78.60^{d}
		Seed	6.37^{l}	2.50^{m}	4.09^{i}	0.23^{d}	86.81^{b}
	Freeze Drying	Peel	6.43^{l}	8.36^{e}	7.57^{e}	0.05^{f}	59.00^{e}
		Flesh	14.99^{g}	7.73^{c}	4.88^{g}	0.04^{f}	72.35^{f}
		Seed	5.00^{m}	3.90^{j}	21.28^{a}	0.05^{f}	69.77^{g}
Buttercup	Raw	Peel	78.27^{d}	4.26^{i}	1.89^{k}	0.33^{b}	15.25^{m}
		Flesh	84.86^{c}	4.69^{g}	0.48^{m}	0.08^{f}	9.89^{l}
		Seed	40.96^{f}	3.67^{k}	11.93^{d}	0.07^{f}	43.37^{h}
	Cabinet Drying	Peel	7.39^{k}	5.12^{f}	6.95^{f}	0.03^{f}	80.50^{c}
		Flesh	9.44^{j}	5.50^{e}	4.56^{h}	0.03^{f}	80.47^{c}
		Seed	4.32^{n}	3.02^{l}	19.26^{c}	0.53^{a}	72.87^{f}
	Freeze Drying	Peel	2.63^{o}	3.63^{k}	6.97^{f}	0.32^{b}	86.45^{b}
		Flesh	4.91^{m}	2.96^{l}	3.08^{j}	0.06^{f}	88.99^{a}
		Seed	4.53^{n}	4.48^{h}	20.94^{b}	0.17^{e}	69.88^{g}

^{a,b,c,d,e,f,g,h,i,j,k,l,m,n,o} Means in a column with different superscripts differ ($P<0.01$)

2.2.2 Physical Characteristics of Pumpkin With Corn Grits Breakfast Cereals. The expansion property of extruded determined the structure of final products. An increase of the percentage of pumpkin flour in the breakfast cereals resulted in a reduced product diameter and hence a decrease in the expansion ratio, therefore it's become harder. However, inclusion of pumpkin flour at 10% level was similar to the control (Figure 1). Figure 2 presents the bulk density of products. There was a general trend that as the level of pumpkin flour increased (for all fractions and all percentage levels) the extrudates exhibited higher density. This result is in line with previous research.[4]

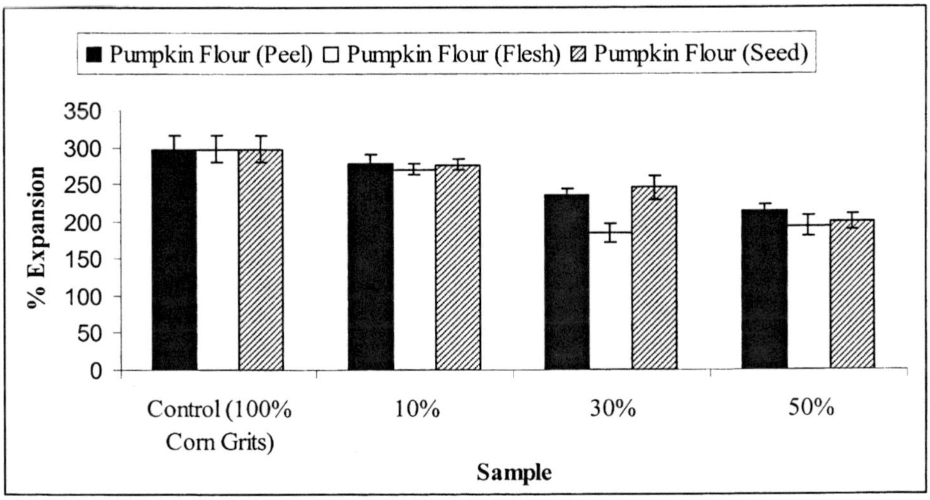

Figure 1 *The expansion ratio of breakfast cereals made from different fractions and percentage of pumpkin flour with corn grits*

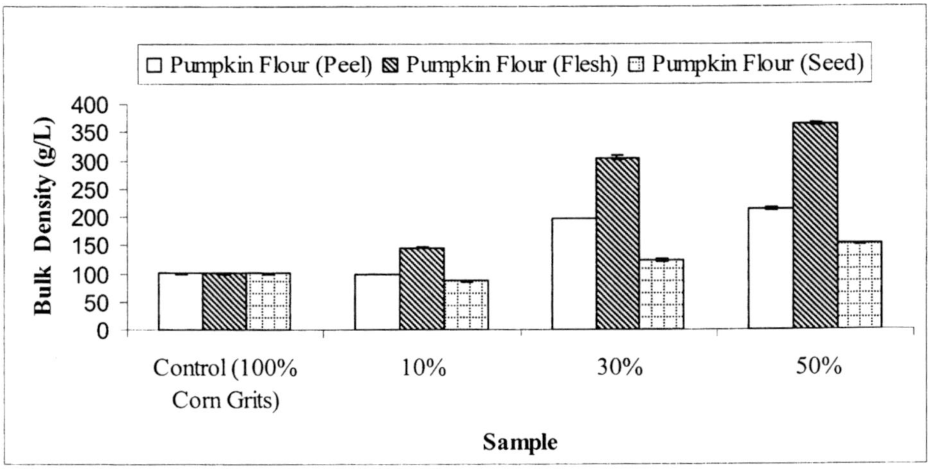

Figure 2 *The bulk density of breakfast cereals made from different fractions and percentage of pumpkin flour with corn grits*

3 CONCLUSION

Results show it is possible to produce pumpkin flour from three (3) fractions (peel, flesh, seed) of pumpkin. This has the potential to increase future levels of utilization of pumpkin. Incorporation of pumpkin flour with corn grits at a lower percentage levels (20% or less) can produce product which similar properties to control sample in terms of the product expansion and bulk density. However this needs further research to carry on in order to get an acceptable extruded product.

References

1 H. Normah and J. Pongjata, *Mal J Nutr.*, 2000, **6**, 65.
2 M.A. Brennan, I. Merts, J. Monro, J. Woolnough and C.S. Brennan, *Starch/Starke.*, 2008, **60**, 248.
3 AOAC, 18th ed., Arlington, Va: Assoc of Official Analytical Chemists Intl. 2005
4 A. Altan, K.L. McCarthy and M. Maskan, *J. Food Eng.*, 2008, **84**, 231.

PEROXIDASE AND LACCASE ACTIVITY AS TOOLS TO CONTROL CROSSLINKING IN ARABINOXYLANS

J.A. Robertson, C.B. Faulds and K.W. Waldron

Sustainability of the Food Chain, Institute of Food Research, Norwich Research Park, Colney, Norwich NR4 7UA, UK

1 INTRODUCTION

Feruloylated polysaccharides have a recognised ability to be oxidatively crosslinked by oxido-reductases such as peroxidases (EC 1.11.1.7). Laccases, (EC 1.10.3.2), more associated with lignin degradation, can also be used for oxidative crosslinking. Cereal residues, like wheat bran (WB) and brewers' spent grain (BSG) contain relatively high concentrations of cell wall bound phenolic acids[1,2] and thus are potential substrates for peroxidase and laccase activity. For WB the concentration of phenolic acids is ~7.6mg.g^{-1} and for BSG the concentration is ~9.2mg.g^{-1}. The bulk of the esterified ferulate (FA) present is apparently unavailable for crosslinking *in muro.*[1] The ability to release feruloylated AX from the cell wall would provide a valuable industrial feedstock for the development of potential texturising agents in foods. However, the potential to enzymatically release AX from cereal residues is limited and chemical extraction results in an extensive de-esterification of AX. This may appear to limit the use of AX as crosslinking agents unless crosslinking can be demonstrated for sparingly feruloylated AX. Using commercial sources of such AX their ability to be crosslinked by commercially available peroxidases and laccases activity has been screened.

2 MATERIALS AND METHODS

2.1 Material

A Sugar Beet pectin extract (SBP) (Herbstreith & Fox KG, Neuenburg, Germany) was used a standard reference gelling material. Water-soluble AX, obtained from Megazyme, Bray, Ireland, as wheat arabinoxylans (WAX), medium viscosity (20cSt) and low viscosity (2cSt) and rye arabinoxylan (RAX) were the test sparingly feruloylated test substrates.
Horseradish peroxidise (HRP) (Sigma, P-8125) had a stated activity of 113 purpurogallin units/mg, AbLAC (Agaricus bisporus laccase) (Sigma 40452) had a stated activity of 7.44U/mg and TvLAC (Trametes versicolor laccase) (Sigma 53739) had a stated activity of 22.4U/mg. Each enzyme was used as supplied, without further purification. Hydrogen peroxide (H_2O_2) was obtained as a 50% solution in water (Sigma-Aldrich: 51,681-3) and diluted to 5% prior to use. All other chemicals used were at least Analar quality.

2.2 Methods

2.2.1 Oxidative crosslinking. For crosslinking reaction, the SBP extract was prepared at known concentrations between 10 -100mg/mL in deionised water. AX extracts were prepared at 25mg/mL in deionised water. Enzymes were diluted in deionised water to an activity: ~ 0.11 U/mL prior to use. Hydrogen peroxide was prepared at 10μl/mL for reaction with HRP but was absent, not required, for gelling reactions using laccase. Gelling reactions were done at room temperature, (20^0C), with time monitored from the addition of enzyme.

Strong gelling was recorded as the development of a firm gel within 5min and good gelling, the development of a firm gel within 15min. Poor gelling was recorded as no firm gel produced within 20min, though 'solution' viscosity increased and no gelling was indicated by the persistence of a solution and no apparent increase in viscosity. All samples were left overnight (16h) to allow equilibration of crosslinking reactions but in practice there was little change in gelling outcome beyond that recorded during the first 15min of reaction.

2.2.2 Phenolic acid analysis. The total alkali-extractable phenolic content of samples was determined following saponification of samples with 4M NaOH (~5mg.mL^{-1}) for 18 h in the dark at room temperature and with gentle stirring and constituent phenolic acids were separated using HPLC (LUNA C18 reverse phase HPLC column (Phenonomex, Macclesfield, UK) and quantified according to the method of Waldron et al. (1996),[3] using *trans*-cinnamic acid as standard.

2.2.3 Sugar analysis. Sugars were released from the samples through hydrolysis in 1M H_2SO_4 for 1h at 105^0C^2. Hydrolysates were derivatised as their alditol acetates, and analysed by GLC using a flame ionisation detector.[4] Total uronic acid content was determined spectrophotometrically using galacturonic acid as a standard.[5]

3 RESULTS

Uronic acid (420mg.g^{-1}) and arabinose (192mg.g^{-1}) were the only sugars detected in significant amount in the SBP fraction, consistent with the presence of feruloylated arabinans.[6] The ara : FA ratio was ~25:1 in SBP. In the WAX and RAX extracts arabinose and xylose accounted for >95% of the sugars present and for each extract the ara:xyl ratio was ~40:60. This ratio is consistent with the ratio for endosperm-derived AX found in water-soluble extracts of wheat and rye.[7] The ratio ara:xyl also suggests an apparently similar degree of substitution in the WAX and RAX. Ara : xyl : FA ratios were: WAX medium viscosity, 590:840:1; RAX, 340:530:1. The low viscosity WAX had no detectable FA. Differences in ester substitution between pectins and AX,[8,9] structural differences in the arabinose chain and distribution of FA substituents can be proposed to affect the mobility and availability of FA for cross linking.

The sugar beet pectin was gelled equally well using either HRP or Tv LAC but gelling was limited or absent using Ab LAC for all substrates (Table 1) even though the active sites are similar for Ab LAC and Tv LAC.[10]

WAX low viscosity was not gelled by any enzyme used, whilst RAX and WAX medium viscosity were gelled using either HRP or Tv LAC.

Table 1. *Isolated polysaccharide gelling characteristics*

Sample	Sugar Beet Pectin	WAX med viscosity	WAX low viscosity	Rye AX
HRP	+++	++	-	+++
Ab LAC	+?	-	-	+?
Tv LAC	+++	++	-	++

+++, gelled <5min; ++, gelled 5-15min; +?, gelled poorly (>20min); -, no gelling

Ferulic acid was the only phenolic acid detected in untreated samples (Figure 1). Following gelation treatment a range of diferulates were also detected. Although ferulic acid concentration in AX was <20 than that in the pectin, (Table 2), gelling was apparently equivalent.

Table 2. *Total Extractable Phenolics: pre- and post-gelling (mg/g)*

Enzyme	*Pectin*	*WAX medium viscosity*	*RAX*
Untreated	7.7	1.1	1.6
HRP	4.4	0.5	0.5
Ab LAC	5.2	1.1	0.3
Tv LAC	3.8	0.5	0.2

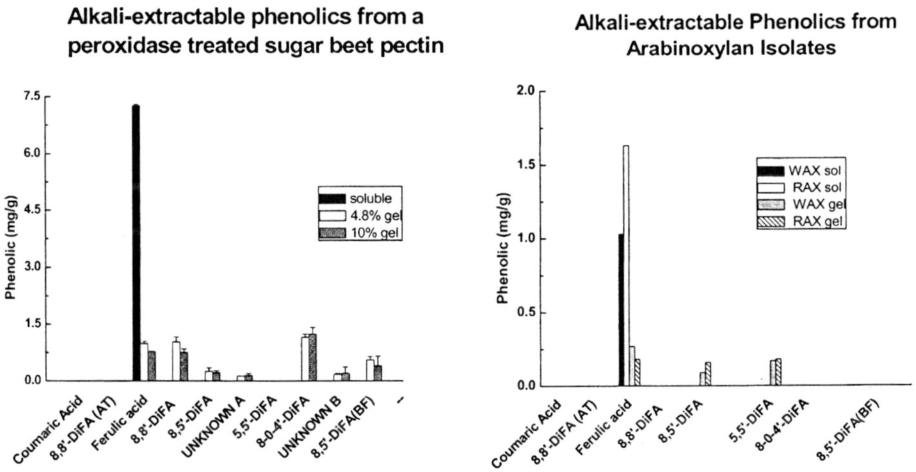

Figure 1. *Alkali-Extractable Phenolic Acids Pre- and Post Gelling Treatment*

The apparent loss in FA and total extractable phenolic acids (Table 2) and appearance of diferulates confirms cross linking had occurred and also includes higher MW phenolic acid complex formation.

The profile of diferulates varied between AX and pectin (Figure 1) and this may be related to polymer structure differences, with FA esterified through an O-5 linkage to arabinose in AX, whilst in pectic polysaccharides the O-2 linkage through arabinose predominates.[8,9]

4 DISCUSSION

Peroxidases and laccases can effectively crosslink AX from cereals even when FA is present in very low concentration. This gives the potential to modify mechanical properties through controlled enzyme activity at relatively low polysaccharide concentration. Also, a relatively low degree of substitution should help overcome apparent steric restrictions on crosslinking.[1] However, as shown using laccases, crosslinking ability can be enzyme source specific. This may reflect crosslinking being a secondary activity to the lignin degradation activity of laccases. The differences in diferulate profile between substrates probably also indicates that chemical structure and polymer conformation and the propensity for molecular interaction are important considerations for designing functional properties.

5 CONCLUSIONS

Oxido-reductases, like peroxidases and laccases, have the potential to modify oxidative crosslinking between phenolic acids at relatively low degrees of phenolic acid substitution on AX. Different enzyme sources, with apparently similar active sites, vary in their ability to promote crosslinking. Polymer structure may also affect the profile of crosslinking.

The ability to modify crosslinking in AX with relatively low degrees of phenolic acid substitution should give the ability to modify and control the mechanical properties of potential texturising agents in foods.

Acknowledgements

We gratefully acknowledge funding from the Biotechnology and Biological Sciences Research Council (BBSRC), UK, and the Commission of the European Communities contract FOOD-CT-2005-006922 "Reducing Food Processing Waste". This study does not necessarily reflect the views of the commission and its future policy in this area.

References

1 J.A. Robertson, C.B. Faulds, A.C. Smith and K.W. Waldron. *J. Agric. Food Chem.* 2008, **56**, 1720.
2 G. Mandalari, C.B. Faulds, A.I. Sancho, A. Saija, G. Bisignano, R. LoCurto and K.W. Waldron. *J. Cereal Sci.* 2005, **42**, 205.
3 K.W. Waldron, A.J. Parr, A. Ng and J. Ralph. *Phytochem. Anal.* 1996, **7**, 305.
4 A.B. Blakeney, P.J. Harris, R.J. Henry and B.A. Stone. *Carbohydr. Res.* 1983, **113**, 291.
5 N. Blumenkranz and G. Asboe-Hansen. *Anal. Biochem.* 1973, **54**, 484.
6 F. Guillon, J-F. Thibault, F-M. Rombouts, A.G.J. Voragen and W. Pilnik. *Carbohydr. Res.*, 1989, **190**, 97.

7 G. Dervilly-Pinel, L. Rimsten, L. Saulnier, R. Andersson and P. Åman, *J. Cereal Sci.* 2001, **34**, 207.
8 A. Oosterveld, I.E. Pol, G. Beldman, A.G.J. Voragen. *Carbohydr. Polym.*, 2001, **44**, 9.
9 M.M. Smith and R.D. Hartley. *Carbohydr. Res.,* 1983, **11**, 65.
10 www.expasy.org/

EFFECTS OF STORAGE AND ASSOCIATED PROCESSING ACTIVITIES ON TEXTURE, STRUCTURE AND MICROBIOLOGY IN NOVEL INGREDIENTS FROM AGRI-BASED SUSTAINABLE SOURCES

J. Robertson,[1] S. Collins,[1] K. Maltby,[1] G.K. Moates,[1] J. Newman,[2] E. Saggers,[2] T. Brocklehurst,[2] K. Wellner[3] and K.W. Waldron[1]

[1]Sustainability of the Food Chain, [2]Microbial Ecology, [3]Imaging, Institute of Food Research, Norwich Research Park, Colney, Norwich NR4 7UA, UK

1 INTRODUCTION

The plant cell wall is an important component in foods and is involved in the control of texture. Texture is important for food acceptability. There are opportunities to process agri-based crops as ingredients to develop naturally structured products with improved stability for transport and storage. Processing activities for accepted agri-based crops can also be extended to valorise nutrient-rich co-products, such as broccoli stems, currently lost as waste materials. Thus the environmental impact of the waste is reduced and the sustainable management of organic matter from food production and processing is enhanced. There is also the potential to contribute to food quality and safety. The effects of storage and downstream food processing on the texture and structural characteristics of carrot and broccoli tissues have been measured together with monitoring the tissue-associated microbial flora.

2 MATERIALS AND METHODS

2.1 Material

Carrots (Nairobi) were sourced from a local farm, transferred from field to laboratory within 24h and stored for up to 112 days at 0 – 5 °C; 90-98% RH . Broccoli (Monaco) was grown under controlled agronomic conditions. Floret heads were harvested and transferred to storage within 24h. Heads were stored for up to 21 days at 50 °C; 90-98% RH.

2.2 Methods

2.2.1 Microbiology. Five individual broccoli heads and carrots were aseptically dissected into component tissues. The component tissues from all 5 samples were pooled and the total aerobic mesophilic bacterial population for each component enumerated at various time points throughout the storage period.

2.2.2 Texture analysis. Texture was analysed using a TA.XT2 Texture Analyser (Stable Micro Systems Ltd, Godalming, Surrey, UK). For carrot and broccoli stem, discs (~1cm thick) were tested using a 5mm diameter spherical probe with 30 kg load cell to measure peak force penetration.

The plant cell wall is an important component in foods and is involved in the control of texture. Texture is important for food acceptability. There are opportunities to process agri-based crops as ingredients to develop naturally structured products with improved stability for transport and storage. Processing activities for accepted agri-based crops can also be extended to valorise nutrient-rich co-products, such as broccoli stems, currently lost as waste materials. Thus the environmental impact of the waste is reduced and the sustainable management of organic matter from food production and processing is enhanced. There is also the potential to contribute to food quality and safety. The effects of storage and downstream food processing on the texture and structural characteristics of carrot and broccoli tissues have been measured together with monitoring the tissue-associated microbial flora.

3 RESULTS

All samples studied show microbial colonisation, up to $\sim 10^6$ organisms / g fresh weight: this is tissue specific but not greatly affected by storage. Carrot (Fig. 1) shows greater colonisation in phloem (outer) tissues. This is not due to the presence of carrot skin / epidermis. Colonisation of broccoli tissues is more variable. Floret tissues (Fig. 2) tend to have higher levels of colonisation, whilst stem core shows little colonisation. Surface tissues (stem epidermis) and exposed cut surfaces (stem end) are susceptible to colonisation. Overall, the levels of microbial contamination of vegetative tissues are within the levels expected for these types of products and remain within this range in samples stored under conditions controlled to preserve sample quality and extend shelf life.

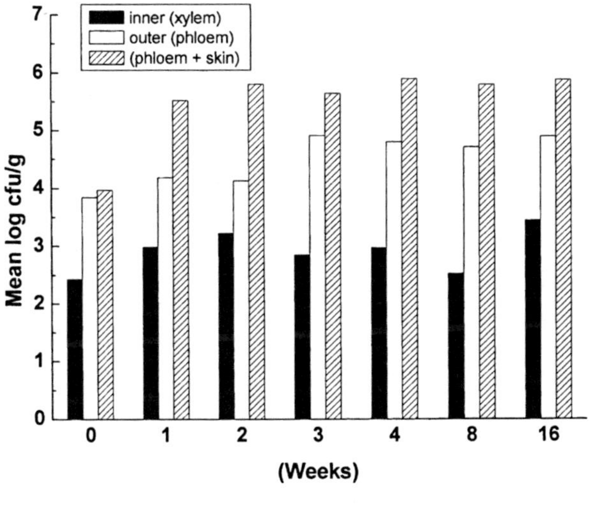

Figure 1 *Microbial colonisation of carrot during storage*

Processing has more effect on carrot texture than storage (Fig. 3). During storage the carrot texture measured does not change. Cooking results in a tissue softening, most noticeable after boiling. Pre-cooking results in a slight tissue softening but pre-cooking prior to boiling remains similar to boiling. Broccoli tissues also show no significant change during storage but soften on boiling (Fig. 4). The vascularised outer stem tissues are toughest but there is little difference between stem core, side stem and floret (not shown) tissues. The results show that broccoli can be stored to extend its shelf life for processing and that its constituent tissues are all suitable for use in processed products.

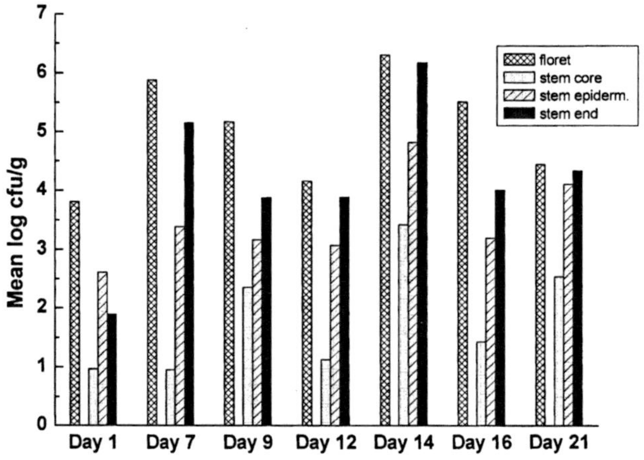

Figure 2 *Microbial colonisation of broccoli during storage*

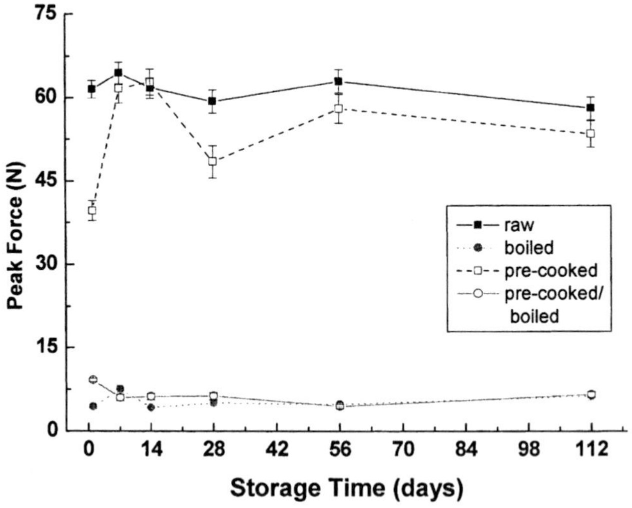

Figure 3 *Effect of Storage and Processing on Carrot Texture - Peak Force Penetration (N)*

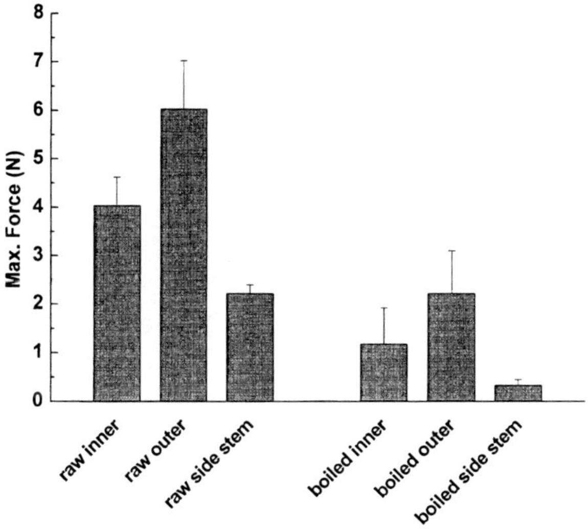

Figure 4 *Effect of Cooking on Broccoli Tissue Texture - Peak Force Penetration (N)*

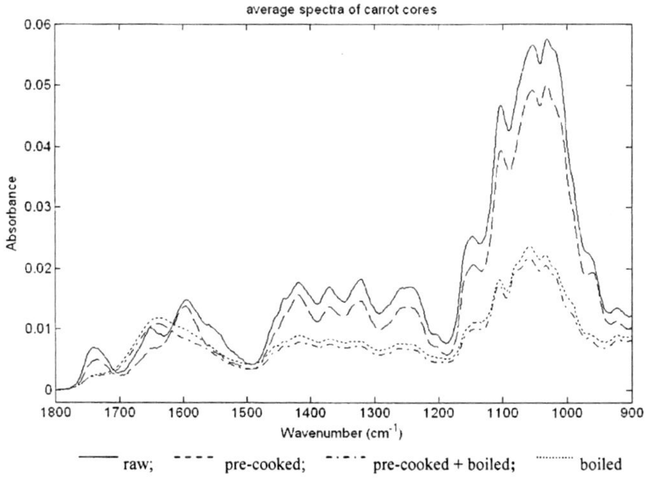

Figure 5 *FTIR spectra of raw and processed carrot*

FTIR spectra (Fig. 5) showed that cooking (boiling) results in structural changes within tissues; these correlate with the change in texture through cooking and relate mainly to pectic polysaccharides. Storage results in no significant change to the FTIR spectra of tissues (not shown). Changes in cell wall / AIR composition do occur during storage, notably a modification (decrease) to the arabinose and galactose containing side chains of pectic polysaccharides (Fig. 6). The change in glucose during storage is related to starch loss. Composition of AIR does not change through cooking (results not shown), indicating that changes in texture result from structural change rather than component loss during processing. This allows for processing to control texture and to maintain composition in products developed.

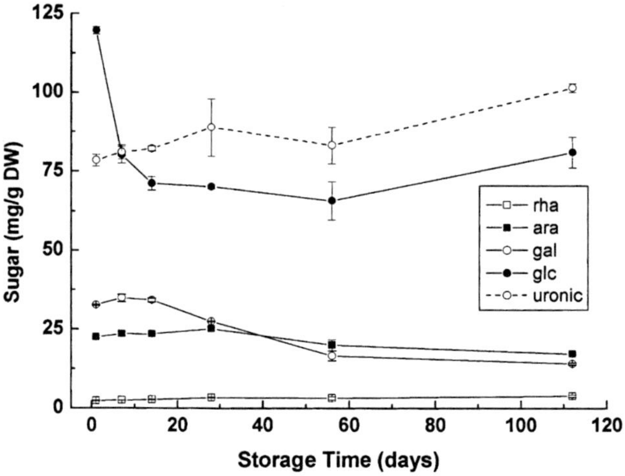

Figure 6 *Effect of Storage and Processing on Component Cell Wall Sugars in Carrot*

4 CONCLUSION

Vegetable tissues remain colonised at expected microbiological levels during controlled storage. Changes in composition / structure do occur during storage but processing has the greater effect on component tissue texture and structure. All tissues can be considered suitable for processing. Through storage, the 'shelf-life' of raw materials can be extended for downstream processing to produce healthy and attractive ready-to-eat products based on fruits and vegetables.

Acknowledgements

We gratefully acknowledge funding from the Biotechnology and Biological Sciences Research Council (BBSRC), UK, and the Commission of the European Communities contract FOOD-2004-T5.4.1.3: Project **023115**. This study does not necessarily reflect the views of the Commission and its future policy in this area.

ANTIOXIDANT PROPERTIES OF *Gracilaria birdiae* AND *Gracilaria cornea*, TWO RED SEAWEEDS FROM THE BRAZILIAN COASTS

B.W.S. Souza, M.A. Cerqueira, J.T. Martins, J.A. Teixeira and A.A. Vicente*

IBB-Institute for Biotechnology and Bioengineering, Center of Biological Engineering, Universidade do Minho, Campus de Gualtar, 4710-057, Braga, Portugal

1 INTRODUCTION

Several synthetic antioxidants such as butylated hydroxyanisole (BHA), butylated hydroxytoluene (BHT), and butylated hydroxyquinone (TBHQ) are commercially available and currently used; however, concerns on their safety and toxicity are hindering their use by the food industry.[1,2,3] Therefore, the development of alternative antioxidants from natural origins has drawn increasing attention, particularly in the most recent years during which the consumer's awareness on food quality and safety issues has improved very significantly. Research efforts have therefore focused on e.g. medicinal plants for the extraction of natural and low-cost antioxidants that can replace synthetic additives.[4]

Many researchers have reported the existence of various types of antioxidants in different kinds of higher plants.[5,6] More recently, reports have revealed seaweeds to be a rich source of antioxidant compounds.[7,8,9] The seaweeds are considered a traditional diet in different regions of the globe, particularly in Asian countries. Their chemical composition shows that they are foods low in calories, with high concentrations of minerals, vitamins and protein, rich in fiber and with relatively high concentrations of poly-unsaturated fatty acids and different antioxidants.[10,11,12]

During their life-cycle, algae are exposed to large amounts of light and high concentrations of oxygen; this combination favors the generation of free radicals, as well as other powerful oxidizers. It is suggested that the absence of oxidative damage in the structural components of the algae and their stability against adverse conditions are due to the presence of antioxidants.[13] Phenolic compounds such as flavonoids, phenolic acids, and tannins are considered to be major contributors to the antioxidant capacity of plants.

Phenolics or polyphenols have received considerable attention because of their physiological functions, including antioxidant, antimutagenic and antitumor activities.[14] Phenolic compounds are commonly found in edible brown, green and red seaweeds, whose antioxidant properties have been correlated to their phenolic content.[15,16] Algal polyphenols, also called phlorotannins, diverge from terrestrial plant polyphenols once the former are derived from phloroglucinol units (1,3,5-trihydroxybenzine). Phlorotannins constitute an extremely heterogeneous group of molecules (with different structures and polymerization degrees) providing a wide range of compounds with potential biological activity.[17]

In that line, this work aims at investigating the antioxidant properties of two different red seaweed extracts from *G. birdiae* and *G. cornea*, collected in the northeast Atlantic coast of Brazil.

2 METHODS AND RESULTS.

2.1 Marine algae collection and preparation of the extracts

Specimens of red seaweeds *Gracilaria birdiae* (*Gb*) were collected on the Atlantic coast of Brazil (Fleixeiras beach – Trairi - Ceará). This species grows attached to rocks or dead coral. The diploid phase that develops directly on the female thallus, the carposporophyte, is evident all year-round in the area, and was selected as seed material. The seedlings were cleaned and then tied in a structure made of string, which was placed in the sea, where it was anchored and submersed for two months. After collection, the algae were cleaned of epiphytes, washed with distilled water and stored at -20 °C.

Specimens of red seaweeds *Gracilaria cornea* (*Gc*) were collected on the Northeast coast of Brazil (Pacheco beach - Ceará). Upon collection, the algae were cleaned of epiphytes, rinsed with tap water and stored a -20 °C before utilization.

A sample of each of the algae (10 g) was extracted with 100 mL of either ethanol (purity 99.8 %, Riedel-de Haën, Germany) or methanol (purity 99.8 %, Riedel-de Haën, Germany) using a soxhlet extractor for 6 h at room temperature. After 6 h, the mixture was filtered, the extract was concentrated in a rotary evaporator at 40 °C and the extraction yields (%) were determined. The extracts were stored at 20° C until further use. All the experiments were conducted in triplicate.

Table 1 *Yield of total extract (expressed as % (w/w) of seaweed on a dry weight basis). Total phenolic content (mg gallic acid equivalents/g extract) of ethanolic and methanolic extracts of Gb and Gc.*

Seaweeds	*Solvent*	*Total phenols*	*yield*
Gracilaria birdiae	Ethanol	1.13 ± 0.03^{a}	4.97 ± 0.52^{a}
	Methanol	1.06 ± 0.07^{a}	4.87 ± 0.66^{a}
Gracilaria cornea	Ethanol	0.88 ± 0.03^{b}	4.92 ± 0.61^{a}
	Methanol	0.89 ± 0.07^{b}	4.72 ± 0.57^{a}

Different superscript letters indicate a statistically significant difference (Tukey test $p < 0.05$).

Solvent extraction is the most common method used in sample preparations from plants. The total yield values of both ethanolic and methanolic extracts of the two red seaweeds under analysis are given in **Table 1**. For the ethanolic extracts, total extraction yields of 4.97 % for *G. birdiae* and 4.92 % for *G. cornea* were obtained. For the methanolic extracts, a slightly higher yield was observed for *G. birdiae* (4.87 %) followed by *G. cornea* (4.72 %). In both cases the results clearly showed that none of the solvents significantly ($p < 0.05$) affected the phenolic content of the extract. This is particularly relevant in the case of ethanol, once the use of this solvent is allowed in the food industry (while e.g. methanol is not). These values are within the range of values found in the literature for seaweeds. Ganesan et al.[16] observed that the methanolic extracts of the red seaweeds *Acanthophora spicifera*, *Euchema kappaphycus* and *Gracilaria edulis* showed yields of 5.01, 3.98 and 2.85 %, respectively. Chandini, Ganesan & Bhaskar [21], studied the *in vitro* antioxidant activities off three Indian brown and observed that among the total methanolic extracts of the seaweeds, *Padina tetrastomatica* had a yield of 12.31% followed by *Turbinaria conoides* (5.76%) and *Sargassum marginatum* (5.45%).

2.2 Analysis of total phenols

The concentration of total phenols in the extracts was measured by a UV spectrophotometer (Varian-UV–VIS Spectrophotometer, Germany), based on a colorimetric oxidation/reduction reaction.[18] The oxidizing agent was the Folin–Ciocalteau reagent (AOAS, 1990). To 0.5 ml of diluted extract (20 mg in 10 ml distilled water), 2.5 ml of Folin–Ciocalteau reagent (diluted 1:10 with distilled water) and 2 ml of Na_2CO_3 (75 g/L) (Riedel-de Haën, Germany) were added. This solution was incubated for 30 min at 25 °C and then cooled. The absorbance was measured at 760 nm. A blank solution was also prepared, where 0.5 ml of distilled water replaced the 0.5 ml of diluted extract and the procedure was repeated for the measurement of the absorbance.

The contents of phenolic acids, in ethanol and methanol extracts of the red seaweeds *G. birdiae* and *G. cornea* are presented in **Table 1**. There was no statistically significant difference ($p < 0.05$) between the total phenolic compounds content for *G. birdiae* when extracted with ethanol (1.13 %) and when extracted with methanol (1.06 %). The same happens with *G. cornea*, which has a maximum content of total phenolic compounds of 0.89 % when methanol was used and of 0.88 % when ethanol was used as extraction solvent. However, there is a significant difference ($p < 0.05$) between the content of total phenolic compounds of *G. birdiae* and *G. cornea*.

Several studies indicate that there is a correlation between the total phenolic compounds content and the radical scavenging capacity.[22] The antioxidant properties of phenolics are a result of their ability to act as reducing agents, hydrogen donors, and free radical quenchers; phenolics can also act as metal chelators therefore preventing the catalytic function of metals in the process of initiating radicals.[23] This is confirmed in the following sections for the two seaweeds considered in the present work.

2.3 DPPH radical-scavenging system

The DPPH-scavenging potential of different extracts was measured, based on the scavenging ability of stable 1.1-diphenyl-2-picrylhydrazyl (DPPH) radicals by *Gb* and *Gc* antioxidants. The ability of extracts to scavenge DPPH radicals was determined by the method of Blois.[19] The absorbance was measured at 517 nm and the results were expressed in terms of the percentage DPPH-scavenging activity relative to the control sample, using the following equation:

$$\%\ Radical\ scavenging\ activity = \left[\frac{(A_{control} - A_{sample})}{A_{control}}\right] x\ 100 \quad \text{Equation 1}$$

Where *A control* represents the absorbance value of the control sample and *A sample* corresponds to the absorbance value of the sample.

The IC_{50} value represents the concentration of the compounds that caused a 50 % inhibition of the radical scavenging activity. All experiments were carried out in triplicate.

DPPH is a compound that possesses a nitrogen free radical and is readily destroyed by a free radical scavenger. This assay was used to test the ability of the antioxidant compounds present in the algal extracts to function as proton radical scavengers or hydrogen donors.[24] The antioxidants react with the stable free radical DPPH (deep violet colour) and convert it to 1,1-diphenyl-2-picryl hydrazine; this reaction is accompanied with the discoloration of the solution.

The radical-scavenging activity of *G. birdiae* and *G. cornea* ethanolic and methanolic extracts has been evaluated at different extract concentrations and the results are illustrated in **Figure. 1**, expressed as percentage reduction of the initial DPPH absorption by the tested extract. This figure also shows that these values are not far from those obtained with an efficient radical-scavenger such as BHT.

The IC_{50} values for the ethanolic extracts of *Gc* and *Gb* were 0.90 and 0.89 mg/ml, respectively. For methanolic extracts, the IC_{50} values of *Gc* and *Gb* were 4.45 and 1.09 mg/ml respectively. For comparison, the IC_{50} of BHT was 0.45 mg/ml.

Siriwardhana, Lee, Kim, Ha, & Jeon [25] and Lu & Foo [26] reported a high correlation between DPPH radical-scavenging activity and total polyphenolics. Components such as low molecular weight polysaccharides, pigments, proteins or peptides, also influence the antioxidant activity.[25] Devi et al.[22] state that it is possible that the antioxidant activity of seaweed extracts can be the result of the existence of phenolic compounds.

Nevertheless, the presented results indicate that these seaweeds, widely distributed in Brazil, exhibited similar scavenging activities than closely-related seaweeds used in other studies.[2] The extracts of *Gb* and *Gc* showed a better radical-scavenging activity than did the extract of *Palmaria palmata* (IC_{50} = 12.5 mg/mL^{-1}) [27] and ethanol and methanol extracts of *Kappaphycus alvarezzi* with (IC_{50} values of 3.03 mg/mL^{-1}).[1] Especially when compared with the values of IC_{50} for many other species and extraction procedures/solvents, the IC_{50} values found for *Gc* and *Gb* are very interesting and point at the possibility of using these extracts as antioxidants e.g. in the food industry.

The Tukey test (α = 0.05) was used to determine any significance of differences between specific means (SigmaStat, trial version, 2003, USA).

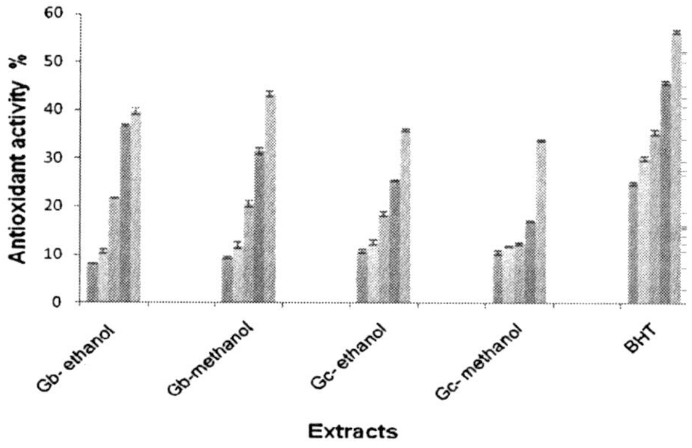

Figure 1. *Antioxidant activities of different solvent extracts of* Gracilaria birdiae (Gb) *and* Gracilaria cornea (Gc) *determined as DPPH radical-scavenging activity. Extract concentration (mg. mL^{-1}): (■) 0.0001; (■) 0.1; (■) 0.5; (■) 0.1; (■) 5. Data are expressed as mean ± standard deviation (n = 3).*

2.4 β-Carotene–linoleic acid assay

The antioxidant activity of extracts was also evaluated by the β-carotene– linoleate model system [20], as described by Shon, Kim, & Sung[6] with some modifications.

The results were expressed in terms of the percentage antioxidant activity relative to the control sample, using the following equation:

$$\%\ Antioxidant\ activity = 100\ x\left[1 - \frac{A_S^0 - A_S^{120}}{A_C^0 - A_C^{120}}\right]$$

Equation 2

Where A_S^0 is the absorbance of sample at time t = 0 min and A_S^{120} is the absorbance of sample at time t = 120 min, and A_C^0 is the absorbance of the control sample at time t = 0 min and A_C^{120} is the absorbance of the control sample at time t = 120 min.

The bleaching mechanism of β-carotene is a free radical mediated event resulting from the formation of hydroperoxides from linoleic acid. In the absence of an antioxidant, β-carotene will undergo rapid discoloration.[28] Linoleic acid will become a free radical with a hydrogen atom abstracted from one of its diallylic methylene groups. The radical formed then attacks the highly unsaturated β-carotene molecules. Therefore, the orange coloured chromophore of β-carotene would be degraded and the results could be monitored spectrophotometrically. The presence of antioxidants in the different extracts can protect the extent of β-carotene bleaching by neutralizing the linoleate-free radical and other free radicals formed in the system.[29] Hydroperoxides formed in this system will be neutralized by the antioxidants from the extracts.

Therefore, the degradation rate of β-carotene depends on the antioxidant activity of the extracts. In our work, two solvents (ethanol and methanol) were used to prepare the *Gb* and *Gc* extracts. In the linoleic emulsion system, the oxidation of β-carotene was effectively inhibited by all extracts of *Gb* and *Gc* in different concentrations (Figure 2), achieving values of β-carotene oxidation inhibition of up to 40 %. Again, these are very good values when compared to other systems described in the literature. A previous study has shown that there is a correlation between the content of total phenolic compounds and β-carotene oxidation.[30] Positive correlations were also established between the presence of phenols and antioxidant activities, suggesting that the observed activities of the two seaweeds may be attributed to their content in phenolic compounds.

The Tukey test (α = 0.05) was used to determine any significance of differences between specific means (SigmaStat, trial version, 2003, USA).

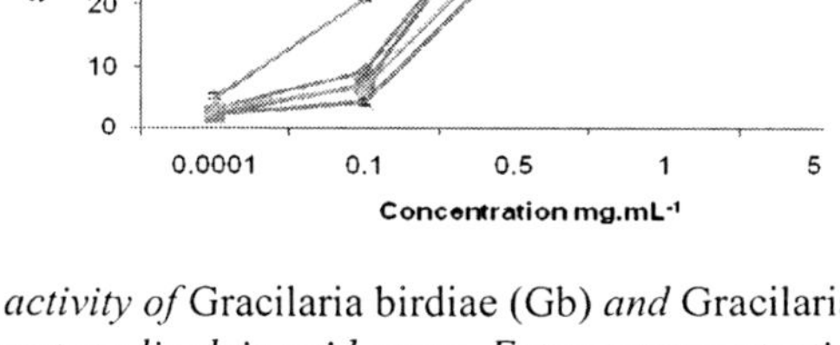

Figure 2 *Antioxidant activity of* Gracilaria birdiae (Gb) *and* Gracilaria cornea (Gc) *samples as determined by β-carotene-linoleic acid assay. Extract concentration (mg. mL^{-1}): (X) Gc-Methanol; (▲) Gb-Methanol; (■) Gc- Ethanol; (♦)G.b-Ethanol; (*) BHT. Data are expressed as mean ± standard deviation (n = 3).*

3 CONCLUSION

Many synthetic antioxidants have shown toxic and mutagenic effects, which have shifted attention towards naturally occurring antioxidants. A great number of naturally occurring substances such as seaweeds have been recognized to have antioxidant properties. In the present work, the solvents extracts of *Gb* and *Gc* exhibited excellent oxygen scavenging effects by DPPH assay and β-carotene linoleic acid assay, while also demonstrating the presence of phenolic compounds in the algae. It can be concluded that these seaweeds can be utilized as a source of natural antioxidants. The findings of the current report appear useful for further research aiming at isolating and identifying the specific phenolic compounds responsible for the antioxidant activity of *Gracilaria cornea* and *Gracilaria birdiae*.

Acknowledgements

The author B.W.S. Souza is the recipient of a fellowship from the Coordenação de Aperfeiçoamento de Pessoal de Nível Superior (CAPES, Brazil), the authors M.A. Cerqueira and J. T. Martins are recipient of fellowships from the Fundação para a Ciência e Tecnologia (FCT, Portugal) (SFRH/BD/23897/2005 and SFRH/BD/32566/2006, respectively). The authors also wish to acknowledge NGO Terramar for donating the algae.

References

1 K.S. Kumar, K. Ganesan and P.V. Subba Rao, *Food Chem.*, 2008, **107**, 289.
2 A. Ismail and T.S. Hong, *Malaysian J Nutr*, 2002, **8**, 167.
3 N. Ito, M. Hirose, S. Fukushima, H. Tsuda, T. Shirai and M. Tatematsu, *Food Chem. Toxicol.*, 1986, **24**, 1071.
4 D. Atmani, N. Chaher, M. Berboucha, K. Ayouni, H. Lounis, H. Boudaoud, N. Debbache and D. Atmani, *Food Chem.*, 2009, **112**, 303.
5 R.A. Larson, *Phytochem.*, 1988, **27**, 969.
6 M.Y. Shon, T.H. Kim and N.J. Sung, *Food Chem.*, 2003, **82**, 593.
7 S.N. Lim, P.C.K. Cheung, V.E.C. Ooi and P.O. Ang, *J. Agric. Food Chem.*, 2002, **50**, 3862.
8 T. Kuda, M. Tsunekawaa, H. Goto and Y. Araki, *J. Food Compos. Anal.*, 2005, **18**, 625.
9 X.J. Duan, W.W. Zhang, X.M. Li and B.G. Wang, *Food Chem.*, 2006, **95**, 37.
10 A. Jiménez-Escrig and C.I. Goni, *Arch. Latinoam. Nutr.*, 1999, **49**, 114.
11 S.Risso, C. Escudero, S. Estevao, M.L. De Portela and M.A. Fajardo, *Arch. Latinoam. Nutr.*, 2003, **53**, 306.
12 K.H.M. Cardozo, T. Guaratini, M.P. Barros, V.R. Falcão, A.P. Tonon, N.P. Lopes, S. Campos, M.A. Torres, A.O. Souza, P. Colepicolo and E. Pinto, *Comp. Biochem. Physiol. C: Pharmacol. Toxicol. Endocrinol*, 2007, **146**, 60.
13 D.Burritt, J. Larkindale and C.L Hurd, *Planta*, 2002, **215**, 829.
14 A.Othman, A. Ismail, N.A. Ghani and I. Adenan, *Food Chem.*, 2007, **100**, 1523.
15 A. Jiménez-Escrig, I. Jiménez-Jiménez, R. Pulido and F. Saura-Calixto, *J. Sci. Food Agric.*, 2001, **81**, 530.
16 P. Ganesan, C.S, Kumar and N. Bhaskar, *Bioresour. Technol.*, 2008, **99**, 2717.
17 P. Burtin, *P. E J E A Fche*, 2003, **2**, 498.

18 M.Skerget, P. Kotnik, M. Hadolin, A. Rizner-Hras, M. Simonic and Z. Knez, *Food Chem.,* 2005, **89**, 191.
19 M.S. Blois, *Nature,* 1958, **181**, 1199.
20 H.E.A. Miller *J. Am. Oil Chem. Soc*, 1971, **48**, 91.
21 S.K. Chandini, P. Ganesan and N. Bhaskar, *Food Chem.,* 2008, **107**, 707.
22 K.P. Devi, N, Suganthy, P. Kesika and S.K. Pandian, S.K. *BMC Complement Altern Med*, 2008 **8**, 38.
23 X.J. Wu and C. Hansen, *J. Food Sci.*, 2008, **73**, M1.
24 N. Singh and P.S. Rajini, *Food Chem.*, 2004, **85**, 611.
25 N. Siriwardhana, K.W. Lee, S.H. Kim, J.W. Ha and Y.J Jeon, *Food Sci. Technol. Int.,* 2003, **9**, 339.
26 Y.R. Lu and L.Y. Foo, *Food Chem.,* 2000, **68**, 81.
27 Y.V. Yuan, M.F. Carrington and N.A. Walsh, *Food Chem. Toxicol.*, 2005, **43**, 1073.
28 G.K. Jayaprakasha, R.P. Singh and K.K. Sakariah, *Food Chem.,* 2001, **73**, 285.
29 Y.L. Chew, Y.Y. Lima, M. Omar and K.S. Khoo, *LWT Food Sci. Technol*, 2008, **41**, 1067.
30 Y.S. Velioglu, G. Mazza, L Gao and B.D. Oomah, *J. Agric. Food Chem*, 1998, **46**, 4113.

Value Added Products – Fish, Meat and Dairy

DAIRY SIDE STREAM VALORISATION

N. Hotrum, M. Fox, C. Akkerman, P. de Kok and P. de Jong

NIZO food rescarch, P.O. Box 20, 6710 BA Ede, The Netherlands

1 INTRODUCTION

Side stream valorisation can directly benefit food manufacturers by reducing waste removal costs, gaining entry to new value-added markets and/or improving sustainability in the food process chain (Table 1). Moreover, food producers are in need of innovative methods to deal with side streams as governments and regulatory authorities continue to tighten environmental policies concerning waste disposal in response to global and community pressure. In general, there are three main ways to approach side stream valorisation (i) stabilization of a co-product intermediate for further processing by e.g. a third party, (ii) identification and isolation of a high value fraction from the side stream and (iii) implementation of strategies to reduce the volume of low value side streams. To select the most appropriate route, an integral approach can be applied. First, the side stream of interest is identified and characterised and components with potential value are identified. Second, potential processing routes to stabilise and/or isolate the desired components are identified and possible markets for the side stream components are identified. This results in a selection of the components/markets with the highest potential. Third, process design and testing taking into account existing and new technologies, process economics and product requirements is performed. Leading, finally, to application testing, scale up and process implementation. This paper presents strategies for how to apply this valorisation approach to the three main valorisation routes indicated above.

Table 1 *Strategies, benefits and timelines for dealing with side streams*

Strategy	**Benefit**			**Timeline**		
	reduce waste removal	value-added markets	improved sustain-ability	quick win	mid-term	long term
Stable intermediate	++	+	+	++	+	+/-
High value fraction	+/-	++	+/-	-	++	+
Reduce volume side stream	++	-	++	+/-	+	+

2 METHOD AND RESULTS

2.1 Creating value from a waste stream: a stable intermediate

Up until the mid twentieth century whey, a co-product of cheese-making and casein manufacture, was considered a waste stream and disposed of as fertilizer, waste water effluents or for a low return as animal feed. However, with the advent of modern separation techniques, the nutrient rich whey stream has developed into a valuable dairy ingredient. The key scientific, technological and sociological milestones that have facilitated this transformation have been elegantly summarized in a recent review.[1]

The total world production of liquid whey is ~ 180 million tonnes per year, of which it is estimated that approximately 60% is applied to high grade food and industrial use; this use being the highest in New Zealand/Australia (90%), followed by North America (80%), European Union (60%) and rest of world (25%). Considering that the remaining 40% (~ 70 million tonnes per year) is used for feed, fertilizer or waste, optimising whey processing is still relevant today. This was demonstrated recently by NIZO food research in collaboration with the Orkney Cheese Company Ltd. a premium Cheddar cheese producer based on the Orkney Islands, north of Scotland. NIZO food research was requested to perform a valorisation scan to determine the most promising approach to utilisation of the premium quality whey released during cheese production waste stream. Several concept processing routes were identified and economically evaluated taking into account transportation and scale issues associated with the remote production location. The valorisation scan identified opportunities to minimize waste disposal and add value to the waste whey stream at the same time. There is a large potential in particular from small to medium scale cheese producers, to further reduce waste stream volumes through the stabilisation of whey.

2.2 Creating value from a waste stream: a high value fraction

Once a stable intermediate is available, the opportunity to further exploit the intermediate through a series of separation and concentration steps arises. In the dairy industry, a multitude of whey fractions (Figure 1) are commercially available.

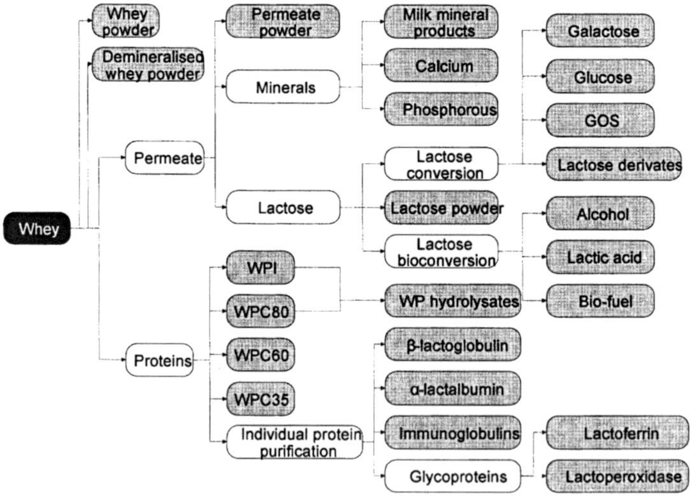

Figure 1 *Example of range of high value fractions that can be isolated from whey*

The generation of high value fractions is rapidly developing in the wider food industry. For example, as many foods contains precious aromas and taste compounds, it is inevitable that waste streams from the production of foods have (residual) concentrations of these desirable and valuable compounds. It has been shown that aromas, sweet- (not sucrose), salt- (not sodium or potassium salts)[2] and umami-enhancing compounds can be found present in these systems.

NIZO food research has developed technology to detect sensory active, food grade, natural ingredients in complex foods and food derived products such as in waste streams.[3] The approach is depicted schematically in Figure 2. This technology uses a cascade of food grade separation steps and highly sensitive sensory panelling in order to determine which fraction contains the valuable sensory enhancing compounds. As the desired compounds are concentrated and purified during this procedure, their activity becomes more pronounced, making it possible to "zoom in" on the desired compound starting from extremely complex food compositions.

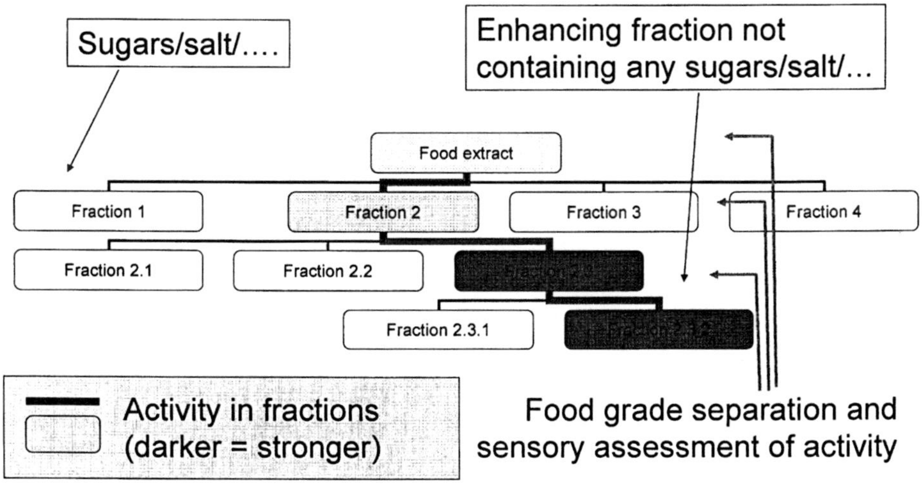

Figure 2 *Approach to detect sensory active, food grade, natural ingredients in waste streams*

For the isolation of valuable sensory compounds, an automated in-line/on-line screening tool is available at NIZO food research incorporating the patented Olfactoscan technology.[4] This technique has been applied to identify and investigate aroma compounds that increase salt perception without adding salt, boost sweetness without adding sugars and enhance taste without the need for monosodium glutamate. In addition, novel masking compounds have been studied to prevent the perception of specific off-flavours. Identification of the desired fraction is the first step in creating a high value fraction. Once the fraction of interest has been identified, a production process targeted at harvesting these compounds can be developed.

2.3 Creating value from a waste stream: reducing waste stream volume

Rinsing fluid from cleaning is an example of a low value waste stream, where reducing volume can add value for the food manufacturer. Cleaning of processing equipment with its associated water use, cleaning chemicals, product loss, and waste water emissions is a significant source of waste stream volume in the dairy industry. In the Netherlands, 11.7 billion kg of milk is processed annually.[5] Using a NIZO food research calculation tool it is estimated that a 50% reduction in cleaning of heat exchangers could lead to a reduction of 1.7×10^5 inhabitant equivalents (IE) of waste stream production, where IE is the biodegradable organic load with biochemical oxygen consumption for a period of five days at 20°C (BZV_5^{20}) of 60g O_2. Combined with additional benefits including reduction in energy consumption, cleaning chemicals and product losses, a 50% reduction in cleaning of heat exchangers has the potential to reduce production costs by €10 million per year. This corresponds to cost savings of approximately k€150/year for a pasteurization line processing 100,000 tonnes of raw milk, thus demonstrating the potential value of reducing waste stream volume through cleaning reduction.

However, cleaning is an essential aspect of dairy processing required maintain process efficiency and ensure product quality and safety by keeping fouling under control. Fouling is the undesired build up of deposits on equipment surfaces during processing. Two main types of fouling can be distinguished in dairy processing: protein/mineral deposit formation and bio-fouling, both of which have been extensively studied.[6,7,8]

Current routes to reduce the costs and environmental impact of cleaning are based on (i) process optimization and (ii) cleaning cycle optimization. In addition, NIZO food research is currently working in collaboration with the Dutch Dairy Association (NZO) to develop a novel technique to prevent fouling whereby the time between cleaning cycles can be significantly extended (iii). This project forms one aspect of a wider initiative of the Dutch Dairy industry to improve processing efficiency within the framework of the MJA-3 energy efficiency agreement with the Dutch government. Each of these three approaches to cleaning reduction is described briefly below.

Computer simulation offers a rapid and cost effective approach to optimising process performance. For example, NIZO Premia is a predictive tool that can be applied to investigate the effect of key control variables (e.g. temperature, residence time, flow rate) on the level of fouling without the need for costly production trials. The fouling model has been validated based on a correlation between protein denaturation in milk and fouling in heat exchangers[9] that has been confirmed by many investigators.[10,11,12] Production run times can be extended by reducing the level of fouling, thereby extending the time between cleaning cycles and contributing to a reduction in the volume of the cleaning waste stream.

Cleaning cycle optimisation is another route to reducing waste stream volumes. NIZO food research has recent experience with in-line optimization of heating and cleaning equipment using OptiCIP+, a cleaning optimization tool commissioned by the Dutch Dairy Association (NZO).[13] Based on production parameters including flow, temperature drop and product composition, the modelling software predicts and monitors the level of fouling in the processing equipment in order to select an optimum cleaning regime at the end of the production run. Moreover, during cleaning, data regarding the level of fouling in the rinsing fluid is applied to signal switching of cleaning steps, to optimise cleaning conditions and optimise cleaning results. The optimization of cleaning procedures with the aid of OptiCIP+ has the advantages of increasing production capacity, reducing the amount of waste water, cleaning chemicals and product losses, leading to overall production cost savings.

Prevention of fouling through disruption of the mechanism of fouling adhesion will enable an improvement in process efficiency and reduction in waste stream volume over and above the presently approaches described above. This is the aim of a current NIZO research initiative, whereby the mechanical action of ultrasound vibration will be applied to prevent both protein/mineral and bio-fouling during heat processing. At lab scale, it has been demonstrated that both biofilm and protein/mineral fouling is diminished by the application of ultrasound.[14] The level of fouling reduction is dependent on the applied ultrasound conditions. A range of 10-70% protein/mineral fouling reduction compared to the reference has been observed depending on frequency, intensity and duration of application of ultrasound. The application of ultrasound to reduce fouling has been confirmed by other investigators. Examples include, the reduction of biofilm formation in continuous flow pilot scale water cooling system,[15] a decreased rate of protein fouling of a heating element submerged in milk in a static, heated vessel in the presence of ultrasound[16] and an increase in the flux during ultrafiltration in the presence of ultrasound treatment, attributed to an increase in the mass transfer coefficient and looser fouling layer on the membrane compared to the reference experiments.[17] The next step in the current research is to scale up the novel fouling prevention technology from lab to pilot scale.

3 CONCLUSIONS

Dairy side stream valorisation offers an opportunity to improve sustainability, secure markets for waste streams and achieve cost savings for dairy manufacturers. With world industrial utilisation of whey estimated at 60%, there is a large potential for the stabilisation and further processing of whey from cheese manufacture. In terms of isolation of high value components, the dairy industry has advanced processing techniques available for whey separation; this approach is rapidly developing in the broader food industry. For low value waste streams, volume reduction can be an attractive solution. NIZO food research is active in the area of waste stream valorisation and waste stream reduction, with dedicated tools including NIZO Premia, OptiCIP+ and Olfactoscan available.

References

1 G.W. Smithers, *Int. Dairy J.*, 2008, **18**, 695.

2 S. Kremer, J. Mojet, R. Shimojo, *J. Food Sci. Sens. Food Qual.*, 2009, **00**, S1.

3 P.M.T. de Kok, *Low sugar, sweet sensation, Novel approaches to enhance sweetness without the application of artificial sweeteners*, 2009, Proceedings IFT Annual Meeting and Food Expo, Anaheim.

4 C. de Jong, A.E.M. Boelrijk, P.M.T. de Kok, K.M.M. Burseg, EP 07110076.2, filing date 12-06-2007.

5 Productschap Zuivel, *Annual report*, 2007, Chapter 3.

6 T.J.M. Jeurnink, *Milk fouling in heat exchangers*, Wageningen University, 1996.

7 P. de Jong, *Modelling and optimization of thermal treatments in the dairy industry*, TU Delft, 1996.

8 P. de Jong, S. Bouman, L.P.M., H.J.L.J. van der Linden, *J. Soc. Dairy Technol.*, 1992, **53**, 3.

9 P. de Jong, *Trends Food Sci. Technol.*, 1997, **8**, 401.

10 M. Lalande, J.P. Tissier, G. Corrieu, *Biotechnol. Prog.*, 1984, **1**, 131.

11 J.E. Schraml, H.G. Kessler, *Milchwissenschaft*, 1996, **51**, 151.

12 P.J. Fryer, *J. Soc. Dairy Technol.*,1989, **45**, 23.

13 D. Allersma, F. Smit, P. de Jong, *VMT*, 2008, **5**, 28.

14 N.E. Hotrum and M. Fox, unpublished results.

15 T.R. Bott, *Trans IChemE, Part A*, 2001, **79**, 484.

16 S.X.Q. Lin and X.D. Chen, *Trans IChemE, Part C*, 2007, **85**, 57.

17 S. Muthukumaran, S.E. Kentish, M. Ashokkumar, G.W. Stevens, *J. Membr. Sci.*, 2005, **258**, 106.

THE USE OF COMMERCIAL ENZYMES FOR THE PRODUCTION OF POTENTIAL BIOACTIVE PEPTIDES FROM LOW VALUE BOVINE MUSCLE AND BOVINE OFFAL PROTEINS

R. Di Bernardini,[1] P. Harnedy,[1] D. J. Bolton,[1] J.P. Kerry,[2] E.E. O'Neill[2] and A.M. Mullen[1]

[1] Ashtown Food Research Centre, Teagasc, Ashtown, Dublin 15
[2] Departments of Food & Nutritional Sciences, University College Cork, Cork

1 INTRODUCTION

In recent years much attention has been placed on the extraction of novel bioactive components from meat waste streams or processing by-products. During meat processing many waste 'fifth-quarter' by-products and low-value cuts of meat are produced. Many of these by-products and low-value cuts are rich sources of protein and represent good candidate raw materials for the exploration of biologically active peptides. During meat ageing the enzymatic changes affect the protein structures increasing free amino acids and peptides through hydrolysis.[1]

Proteolytic degradation of meat proteins using exogenous enzymes can also yield protein fragments of varying size. Bioactive peptides are in general short peptides about 2-20 amino acids in length and are inactive within the sequence of the parent proteins and only become active when released *in vivo* by enzymes during gastrointestinal digestion, *in vitro* during food processing or by induced hydrolysis. Bioactive peptides have been extracted from different sources and once released, have been shown to exhibit a number of biological activities, some of which include antihypertensive, opioid agonists or antagonists, immunomodulating, antithrombotic, antioxidative, anticancer and antimicrobial.[2] Moreover, peptides with multifunctional properties have also been reported.[3] As a result of their therapeutic properties for treatment or prevention of disease, bioactive peptides have the potential to be used as nutraceuticals in functional foods.[2] Functional foods can provide diseases-preventing and health-promoting benefits beyond satisfying the customers' nutritional needs.[4] In the meat sector peptides with antihypertensive activity have been isolated from chicken, porcine and bovine meat proteins;[5,6,7] antioxidant activity from porcine, chicken, bovine and buffalo meat proteins;[9,10,11,12] antimicrobial activity from porcine spleen and bovine blood proteins[13,14] and opioid activity from bovine blood.[15]

The objective of this study was to monitor the use of two commercial enzymes to hydrolyse bovine by-products/low-value cut of meat to generate protein hydrolysates for subsequent bioactivity screening. Both one-dimensional gel electrophoresis and MALDI ToF mass spectrometry were employed to monitor the hydrolyse process. Furthermore, these hydrolysates were tested for antioxidant and antimicrobial activities.

2 METHODS AND RESULTS

2.1 Extraction cytoplasmic/sarcoplasmic proteins

All bovine muscle and organ tissues, brisket (two aged time periods), lung and liver were obtained from the research abattoir (Meat Industrial Development Unit) at Ashtown Food Research Centre, Teagasc. Liver and lung samples were collected on the day of slaughter, while brisket samples were collected at two different *post-mortem* time points. Extraction of sarcoplasmic/cytoplasmic proteins from all tissue samples were carried out in accordance with the method of Jang and Lee[7] using a low salt buffer pH 7.4. The protein content of each fraction was determined by the Bradford protein assay kit (Bio-Rad) and characterised by 1D SDS PAGE (based on the method by Laemmli[8]). In figure 1 a representative profile of water soluble proteins from liver, lung, brisket (1st and 2nd *post-mortem* time points) is shown.

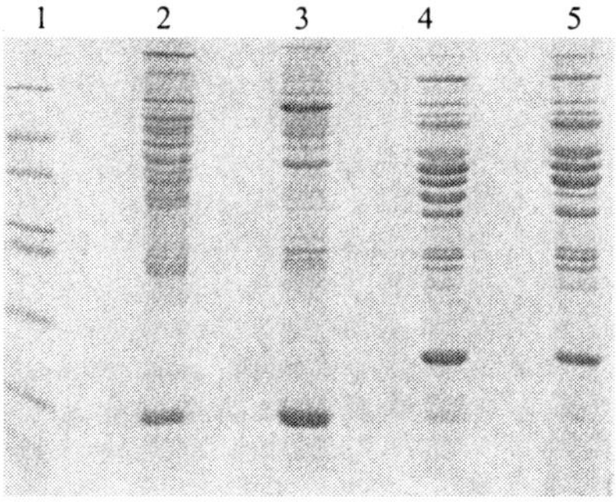

Figure 1. *1D SDS PAGE profiles of liver cytoplasmic proteins, lung and brisket (1st and 2nd post-mortem time points).*
1. Low molecular weight markers (66-6.5KDa); 2. Liver cytoplasmic protein extract; 3. Lung sarcoplasmic protein extract; 4. Brisket (1st *post-mortem* time point) sarcoplasmic protein extract; 5. Brisket (2nd *post-mortem* time point) sarcoplasmic protein extract.

2.2 Hydrolysis with two commercial enzymes

Each protein fraction was hydrolysed with 2 commercial enzymes at seven time periods in triplicate. One-dimensional gel electrophoresis and MALDI Tof Mass Spectrometry were used to monitor the progress and repeatability of each enzymatic hydrolysis. 1D SDS PAGE was used in order to monitor the formation of proteins in the range between 205 and 6.5 KDa. The MALDI Tof was used to monitor the reproducibility of the hydrolysis method of the small sized peptides produced during the hydrolysis, which would not be visible with the 1D SDS PAGE technique.

2.2.1 Monitoring of hydrolysis with 1D SDS PAGE and MALDI ToF mass spectrometry. The repeatability of the seven hydrolysis time points of the three digestions of cytoplasmic/sarcoplasmic proteins from liver, lung and brisket (1st and 2nd *post-mortem* time points) with two commercial enzymes was first monitored by 1D SDS PAGE in triplicate. For each hydrolysis a 12.5 % resolving gel and 15 % resolving gel were used. Results show a decrease in the higher molecular weight band over the seven time points with an increase in lower molecular weight bands. This trend was present in all the three digestions for each tissue type and it was reproducible between the three hydrolysis. A representative gel is presented in figure 2.

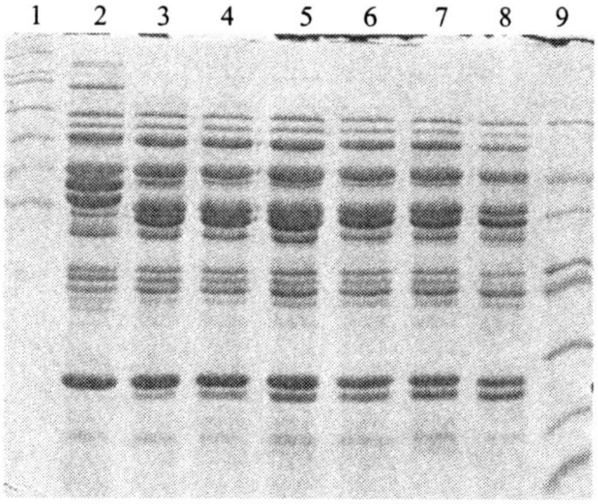

Figure 2. *12.5 % resolving gel 1D SDS PAGE profiles of brisket 2nd post-mortem time point sarcoplasmic proteins digested with the 2nd commercial enzyme at seven hydrolysis time points.*

1. High molecular weight marker proteins (66-6.5 KDa); 2. 1st hydrolysis time point; 3. 2nd hydrolysis time point; 4. 3rd hydrolysis time point; 5. 4th hydrolysis time point; 6. 5th hydrolysis time point; 7. 6th hydrolysis time point; 8. 7th hydrolysis time point. 9. Low molecular weight marker proteins (205-66 KDa).

Liver cytoplasmic proteins, and lung, brisket (1st and 2nd *post-mortem* time points) sarcoplasmic proteins were hydrolysed at two hydrolysis time points in triplicate using two commercial enzymes and analysed by mass spectrometry. Samples were ultrafiltrated with a 10 KDa molecular weight cut off membrane. Ultrafiltrated hydrolysates were studied with MALDI ToF mass spectrometry. The results showed that for the hydrolysis time points of all the tissue types, peptides were formed in the region below 5 KDa, and that the formation of the peptides appeared to be consistent across replicates.

2.2.2 Bioactivity of the hydrolysates. Hydrolysates obtained from each tissue type using 1st commercial enzyme were tested for antioxidant activity with the DPPH (1,1-diphenyl-1-picrylhydrazyl) radical scavenging assay. Each hydrolysis time point was taken in triplicate and the DPPH assay was also carried out in triplicate. Results showed that the hydrolysates from liver, lung and brisket 1st post-mortem time point possessed antioxidant activity. The antioxidant activity of brisket 2nd *post-mortem* time point will be tested. The most interesting tissue(s) or hydrolysis time points will be further purified with RP-HPLC for all the tissue types.

Preliminary study of the antimicrobial activity of bovine by-product protein hydrolysates has been carried out. Liver cytoplasmic proteins were hydrolysed at seven time periods with the 1st commercial enzyme and tested for antimicrobial activity by the agar dilution method. Results showed that fractions had potential antimicrobial activity towards the inhibition of *E.coli* growth, with varying the antimicrobial activity at different hydrolysis time periods.

3 CONCLUSIONS

The hydrolysis methods for the production of bioactive protein hydrolysates and/or peptides from selected bovine low value bovine muscle and bovine by-products have been monitored by one-dimensional gel electrophoresis and MALDI ToF mass spectrometry. 1D SDS PAGE and MALDI methods have provided a means of monitoring the reproducible generation of protein hydrolysates.

Liver, lung, brisket (1st and 2nd *post-mortem* time points) protein hydrolysates generated with one commercial enzyme exhibited antioxidant activity. Furthermore, some of these meat protein hydrolysates have also been found to have potential antimicrobial activity against *E.coli*. Those hydrolysate fractions presenting bioactivity will be purified by RP-HPLC and further characterised.

References

1 G. Koutsidis, J.S. Elmore, M.J. Oruna-Concha, M.M. Campo, J.D. Wood and D.S. Mottram, *Meat Sci.* 2008, **79**, 270.
2 L. Vercruysse, J. Van Camp and G. Smagghie, *J. Agric. Food Chem.,* 2005, **53**, 8106.
3 H. Meisel, *Aust. J. Dairy Tech.,* 2001, **56**, 83.
4 K. Arihara, *Meat Sci.*, 2006, **74**, 219.
5 H. Fujita, K. Yokoyama and M. Yoshikawa, *J. Food Sci.*, 2000, **65**, 564.
6 K. Katayama, H. Fuchu, A. Sakata, S. Kawahara, K. Yamauchi, Y. Kawamura and M. Muguruma, *Asian Australas. J. Anim. Sci.*, 2003, **16**, 417.
7 A. Jang and M. Lee, *Meat Sci.*, 2005, **69**, 653.
8 U.K. Laemmli, *Nature*, 1970, **227**, 680.
9 A. Saiga, S. Tanabe and T. Nishimura, *J. Agric. Food Chem.*, 2003, **51**, 3661.
10 B. Li, F. Chen, X. Wang, B. Ji, Y. Wu, *Food Chem.*, 2007, **102**, 1135.
11 M. Hattori, K. Yamaji-Tsumamoto, H. Kumagai, Y. Feng and K. Takahashi, *J. Agric. Food Chem.*, 1998, **46**, 2167.
12 R. Ohba, T. Deguchi, M. Kishikawa, F. Arsyad, S. Motimura and K. Kida, *Food Sci. Tech. Res.*, 2003, **9**, 91.
13 V. Bonetto, M. Andersson, T. Bergman, R. Sillard, Å. Norberg, V. Mutt and H. Jörnall, *Cell. Mol. Life Sci.*, 1999, **56**, 174.
14 A.C. Fogaca, P.I. Da Silva, M.T.M. Miranda, A.G. Bianchi, A. Miranda, P.E.M. Ribolla and S. Daffre, *J. Biochem.*, 1999, **274**, 25330.
15 J.M. Piot and Q. Zhao, *Biochem. Biophys. Res. Commun.*, 1992, **189**, 101.

BIOACTIVE PROPERTIES OF HYDROLYSATES FROM MACKEREL VISCERA

Z. Khiari, A.B. Martin-Diana, C. Barry-Ryan and D. Rico.

Functional Ingredient Food Unit (FIFU), School of Food Science and Environmental Health, Dublin Institute of Technology (DIT), Marlborough Street, Dublin, Ireland.
Email: zied.khiari@dit.ie

1 INTRODUCTION

More than 50% of the global fish catch is discarded as processing waste which represents a significant cost for the seafood industry. Fish by-products and the under-utilized species are considered low-value resources with negligible market value. However, recent studies have identified a number of bioactive compounds from the fish by-products (i.e. remaining muscle protein, backbone, and collagen). The use of the fish by-products is a new approach to develop more commercial applications. The main objective of this study was to screen and isolate bioactive peptides from Mackerel's viscera hydrolysates produced with different commercial enzymes at different times.

2 METHOD AND RESULTS

2.1 Hydrolysis

Viscera, from Mackerel (*Scomber scombrus*), including all the internal organs (such as kidney, stomach, intestines, pyloric caeca, heart, liver, gonads and gall bladder) were collected and then ground using a meat mincer. The processed viscera were divided in batches and transferred immediately into plastic bags, and stored at -20 °C until further use. Different assays were carried out by varying the time of hydrolysis and the type of enzyme. Times from 2 to 24 hours and different enzymes were used (endogenous enzymes, trypsin, pepsin, chymotrypsin, proteases from *Aspergillus oryzea and* proteases from *Bacillus licheniformis*).

The degree of hydrolysis (DH) was determined using the *o*-phth-aldialdehyde (OPA) method.[1] A significant increment of DH was observed with the time. The degree of hydrolysis increased rapidly in the first 2 hours and then the rate of reaction remained constant with slight increments at the end of the hydrolysis (24 hours).

Significant differences were observed between the enzymes with respect to DH ($p < 0.05$). The highest DH (45%) was achieved after 24 hours by most enzymes except pepsin and proteases from *Aspergillus oryzea* which showed lower DH (22%).

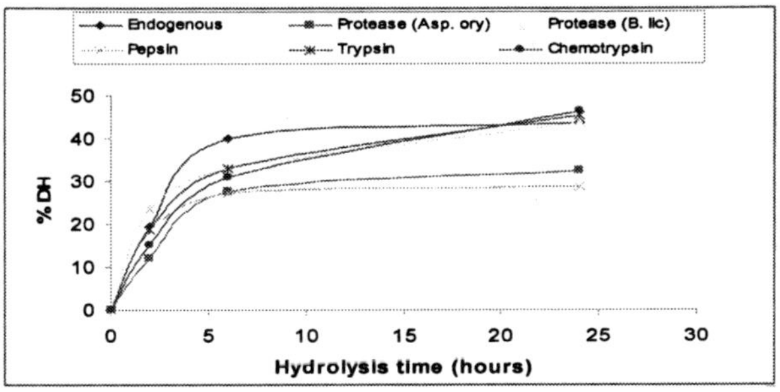

Figure 1 *Degree of hydrolysis (DH) of Mackerel's viscera during hydrolysis with different enzymes. (Asp.ory: proteases from Aspergillus oryzea and B.lic: proteases from Bacillus licheniformis).*

2.2 HPLC

Separation of peptides was achieved by reverse-phase Waters Spherisorb ODS-2 C18 column (150 mm×4.6 mm, 5 μm) on a Waters (Milford, MA, USA) HPLC system equipped with a UV-Vis detector. The mobile phases consisted of A: water and B: acetonitrile (both containing 0.1% TFA), delivered at a constant flow rate of 1ml/min with a linear gradient of 2 - 48% of B over a period of 30min. Elution profiles of Mackerel's viscera hydrolysates separated by reversed-phase HPLC column (ODS-2 C18) are depicted in Figure 2. The majority of compounds (*i.e.,* small peptides or amino acids) were eluted between 2 and 18 min.

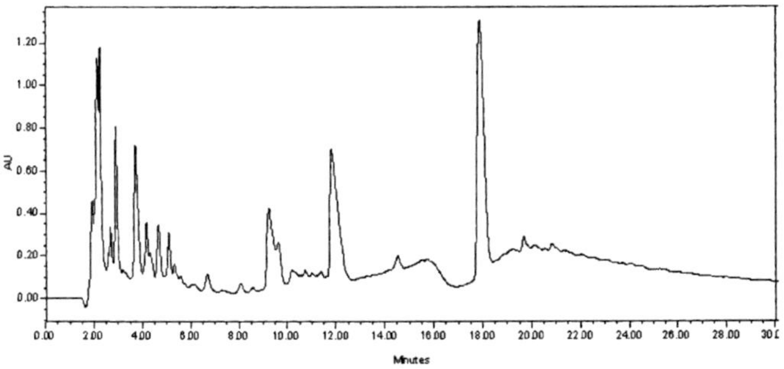

Figure 2 *HPLC chromatogram of hydrolysates from Mackerel viscera after hydrolysis with chymotrypsin for 24 hours.*

Retention of peptides on a reversed-phase column largely depends on the overall hydrophobicity, amino acid composition and ionization state of the peptides.[2] The compounds which eluted earlier in the chromatogram based on the column employed are more hydrophilic than those eluted later.

2.3 Screening for bioactivity

2.3.1 DPPH radical scavenging assay. The scavenging activity was measured using the stable radical *2,2-Diphenyl-1-picrylhydrazyl* (DPPH).[3] 0.5mL of each sample was added to

0.5 mL of 60 μM methanolic solution of DPPH. The mixture was incubated for 30 min in dark at room temperature then the absorbance was measured at 515 nm. DPPH radical-scavenging activity increased with increasing the degree of hydrolysis.
The highest DPPH radical-scavenging activity ($p<0.05$) was achieved with all hydrolysates after 24 hours of hydrolysis. Pepsin was less efficient in term of producing hydrolysates with antioxidant activity on (DPPH) radicals (Figure 3).

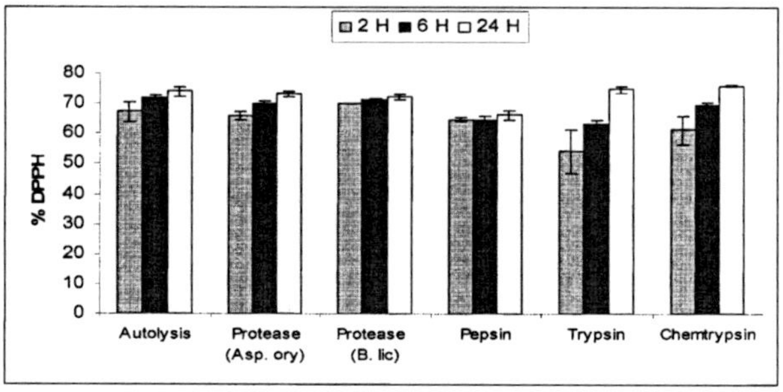

Figure 3 *DPPH radical scavenging activity of Mackerel viscera. Values were given as mean ± standard deviation. (Asp.ory: proteases from Aspergillus oryzea and B.lic: proteases from Bacillus licheniformis).*

2.3.2 Reducing power assay. The reducing power was measured using the ferric reducing antioxidant power (FRAP) assay.[4] A sample (100μl) was mixed with 900 μl of FRAP solution and incubated for 30 min in dark at 37 °C, then the absorbance was read at 595 nm.

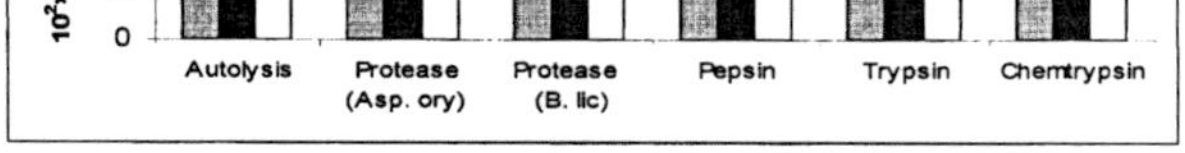

Figure 4 *Reducing power of Mackerel viscera. Values were given as mean ± standard deviation. (Asp.ory: proteases from Aspergillus oryzea and B.lic: proteases from Bacillus licheniformis).*

The reducing power was found to depend on the DH and enzyme used. The reducing power increased in the first 6 hours of hydrolysis then decreased for hydrolysates obtained from the hydrolysis with chymotrypsin, proteases from *Aspergillus oryzea* and autolysis. For the other hydrolysates the reducing power increased as the DH increased. These results were similar to those reported for *yellow stripe trevally* protein hydrolysates.[5]

2.3.3 Semicarbazide-sensitive amine oxidase (SSAO) inhibitory activity (anti-inflammatory activity). The activity of SSAO was determined following the production of hydrogen peroxide at 498 nm, in the presence of 5 mM benzylamine and 10 μg/mL SSAO.[6]

Semicarbazide-sensitive amine oxidase (SSAO) is a common name for a widely distributed enzyme in nature. SSAO plays a key role in inflammation through its catalytic products, hydrogen peroxide, and reactive aldehydes. Therefore, the inhibition of SSAO activity might represent a target for anti-inflammation.[7]

Table 1 *SSAO inhibitory activity.*

Enzymes	SSAO inhibitory activity (%)
Endogenous	16.73±1.08
Protease from *Aspergillus oryzea*	17.28±0.52
Protease from *Bacillus licheniformis*	9.76±0.74
Pepsin	7.12±0.56
Trypsin	30.33±0.11
Chymotrypsin	31.51±0.19

Mackerel viscera hydrolysates, obtained from trypsin and chymotrypsin after 24 hours, showed significant ($p < 0.05$) higher inhibition of SSAO (30.33% and 31.51%, respectively).

3 CONCLUSION

This study shows that the degree of hydrolysis was significantly ($p<0.05$) influenced by the hydrolysis conditions that included time and the type of commercial enzyme used. The antioxidant activity of Mackerel viscera hydrolysates varied with DH and enzyme used. Hydrolysates prepared from Mackerel visceral waste using trypsin and chymotrypsin have potential for applications as good source of peptides with potential antioxidant activity as well as anti-inflammatory compounds by inhibiting the SSAO.

References

1 P.M. Nielsen, D. Petersen and C. Dambmann, *J. Food Sci.*, 2001, **66**, 642.
2 L. Lemieux and J. Amiot, *J. Chromatogr.*, 1989, **473**, 189.
3 G. Yen and J. Wu, *Food Chem.*, 1999, **65**, 375.
4 I.F.F. Benzie and J.J. Strain, *Anal. Biochem.*, 1996, **239**, 70.
5 V. Klompong, S. Benjakul, D. Kantachote, K.D. Hayes and F. Shahidi, *Int. J. Food Sci. Technol.*, 2007, **43**, 1019.
6 A. Holt, D. F. Sharman, G.B. Baker and M.M. Palcic, *Anal. Biochem.*, 1997, **244**, 384.
7 S.Y. Lin, C.C. Wang, Y.L. Lu, W.C. Wu and W.C. Hou, *Food Chem. Toxicol.*, 2008, **46**, 2485.

COMPARISON BETWEEN GELATINES EXTRACTED FROM MACKEREL AND BLUE WHITING HEADS USING DIFFERENT ORGANIC ACIDS

Z. Khiari, C. Barry-Ryan, D. Rico and A.B. Martin-Diana.

Functional Ingredient Food Unit (FIFU), School of Food Science and Environmental Health, Dublin Institute of Technology (DIT), Marlborough Street, Dublin 1, Ireland.
Email: zied.khiari@dit.ie

1 INTRODUCTION

Gelatine represents one of the most important ingredients to modify functional properties and increase the nutritional value of food products. It is mainly used as a gelling as well as an emulsifiers and foam-stabilizing agent. Cattle bones, cattle hides and pork skins are the traditional materials used in gelatine production but for many social-cultural reasons including the possibility of bovine gelatine to transmit infectious illness like bovine spongiform encephalopathy (BSE), alternative sources are increasingly demanded.[1] The fish processing industry generates significant amount of by-products which represent a potential source of gelatine. The main objective of this study was to compare the functional properties of gelatines extracted from Mackerel and Blue Whiting heads using different organic acids: acetic acid (AA), citric acid (CA), lactic acid (LA), tartaric acid (TA) and malic acid (MA).

2 METHOD AND RESULTS

2.1 Gelatine extraction

Gelatine was extracted after treating the minced fish head with NaOH, followed by an acid treatment.[2] Different organic acids, acetic, citric, lactic, tartaric or malic acid were used. After that the material was washed under running water from a tap, and the gelatine was extracted at 42 °C overnight with distilled water.

The yield varied depending on the organic acid used and the raw material from which the gelatine was extracted (Figure 1). For gelatines from Mackerel head, when using citric acid the yield (43.96%) was significantly ($p<0.05$) higher than the rest of gelatines. Gelatine extracted from Blue Whiting head using acetic and citric acid showed significantly ($p<0.05$) the highest yields (44.11% and 45.43%, respectively). While when tartaric acid was used, the yield had the lowest value (28.44%) (Figure 1 (II)).

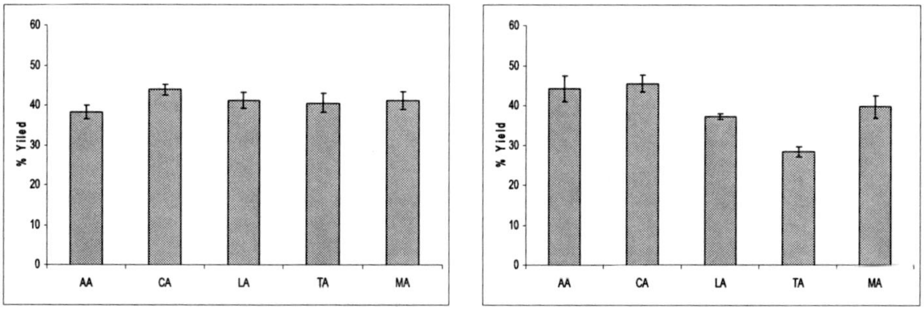

I II

Figure 1 *Yield of gelatine extraction;* ***I:*** *gelatines from Mackerel head,* ***II:*** *gelatine from Blue Whiting head. Values were as mean ± standard deviation.*

The type of acid affects the pH of the extraction solution, which affects the swelling and the breaking of covalent bonds in the collagen cross-links of the raw material.[2] Hence suitable mild acid treatment is essential for gelatine extraction.[3]

2.2 Protein solubility

Protein solubility index was measured by changing the pH from 2.0 to 12.0 of 0.1% gelatine solutions and calculating the protein content of supernatant and the protein content of suspension before centrifugation.[4]

Solubility was significantly ($p<0.05$) affected by the acid used in the extraction. Gelatines from Mackerel extracted with citric acid showed significantly ($p<0.05$) the highest solubility. Gelatine from Blue Whiting did not show any different between citric acid and tartaric where highest values of solubility were observed. All the gelatines regardless of the acid used for the extraction showed similar pH behaviour, highest solubility was observed at acid pH 2 and a decrease in solubility was observed at neutral pH 6-7 maintaining constant until pH 8.

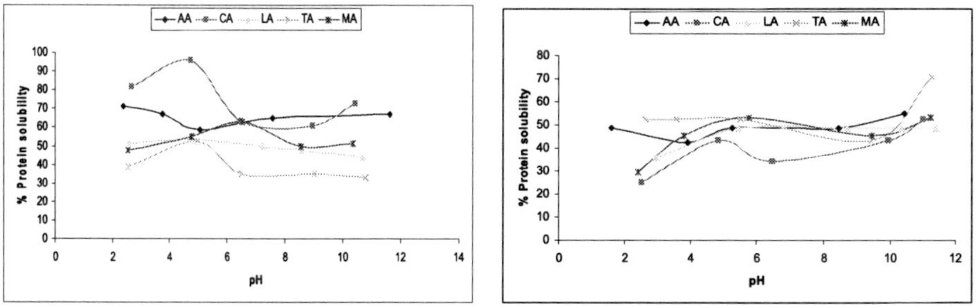

I II

Figure 2 *Protein solubility of Mackerel head gelatines (I) and Blue Whiting head gelatines (II).*

The differences in solubility between Mackerel and Blue Whiting gelatines might result from the differences in amino acid compositions, sequence, molecular weight, conformation and the content of polar and non-polar groups in amino acids.[5]

2.3 Emulsifying properties

Emulsifying properties were determined by the turbidometric method.[6] Emulsions were prepared by adding 2.0 mL of pure sunflower oil to 8.0 mL of gelatine solutions at different protein concentrations (0.05%, 0.1%, and 0.2%) in 50mM phosphate buffer at pH 7.5.

Results showed that increasing concentration of gelatine solution decreased the emulsifying activity (EAI). Also the organic acid used for extraction affected the EAI. Gelatines extracted from Mackerel head using tartaric and acetic acid showed significant ($p<0.05$) highest EAI (27.57 and 27.28%, respectively). No significant difference in stability of emulsion was observed between acids, but the protein content affected the ES, with gelatine extracted using tartaric and citric at 0.1% protein content being the most stable in terms of emulsion stability (Figure 3 (C)).

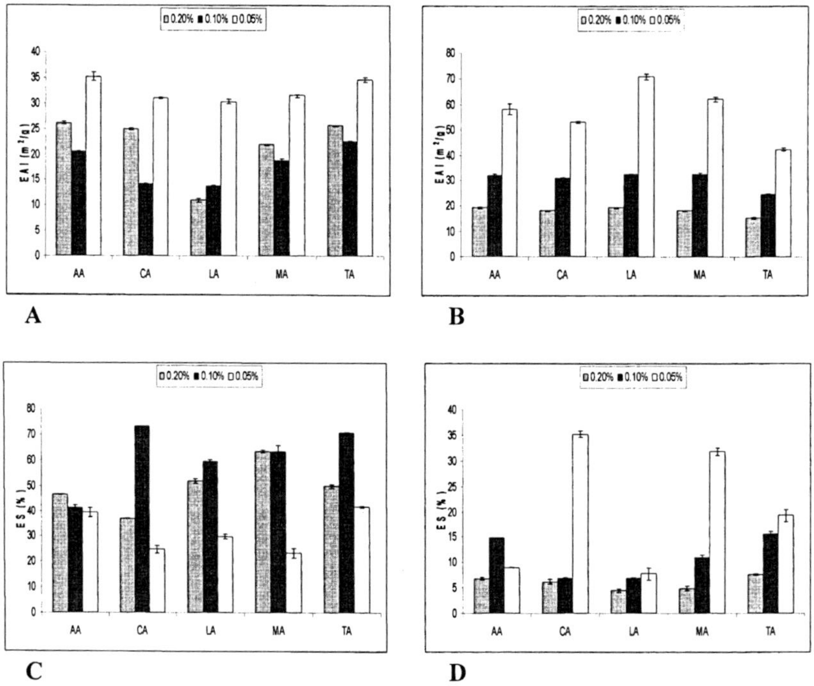

Figure 3 *Emulsion capacity (EAI) of Mackerel head gelatines (A) and Blue whiting head gelatines (B) and emulsion stability (ES) of Mackerel head gelatines (C) and Blue Whiting head gelatines (D). Values were given as mean ± standard deviation.*

Gelatines extracted from Blue Whiting head using lactic acid showed significant ($p<0.05$) high EAI (40.69%) followed by malic and acetic acids (37.54% and 36.49%, respectively). However in terms of stability, lactic and acetic acids derivatives showed significant ($p<0.05$) lowest ES (Figure 3 (D)).

2.4 Foaming properties

Foaming properties, including foaming capacity (FC) and foam stability (FS), were determined using 0.3% gelatine solutions at neutral pH.[7] The foaming capacity (FC) was affected by the organic acid used in the swelling step. Gelatine extracted from Mackerel head treated with tartaric acid showed significant ($p<0.05$) highest FC (56%), followed by acetic (52%) and citric acids (49%). Lactic, tartaric and acetic showed the highest FS (93%). For gelatines extracted from Blue Whiting head, malic, lactic and citric showed FC around 30%, in terms of stability malic acid derived gelatines were significant ($p<0.05$) the most stable emulsions.

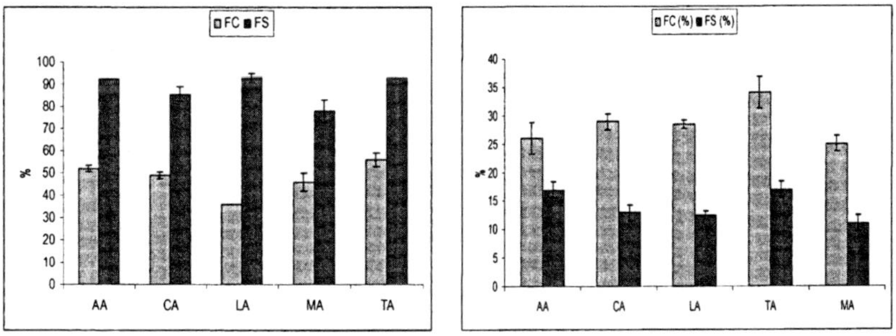

I II

Figure 4 *Foaming capacity (FC) and stability (FS) of Mackerel (I) and Blue Whiting (II) head gelatines. Values were given as mean ± standard deviation.*

2.5 Gel strength

Gel strength (Bloom) was determined using an Instron Universal Testing Machine model 3300 (Instron Ltd., High Wycombe, Bucks, England) with a static load cell of 500 Newtons equipped with a 1.27 cm diameter flat-faced cylindrical probe.

The gel strength of Mackerel and Blue Whiting gelatines was affected by the organic acid. Gelatines from Blue Whiting head treated with tartaric acid showed significant ($p<0.05$) highest bloom (29.57± 1.44g). For gelatines from Mackerel head, malic, citric and acetic extracts showed significant ($p<0.05$) highest bloom (70-75g).

Table 1 *Gel strength of Mackerel and Blue Whiting head gelatines extracted with different acids.*

Treatment	Gel strength (g)	
	Mackerel	**Blue Whiting**
Acetic	70.36±1.44	21.92±2.16
Citric	71.38±7.21	17.84±0.71
Lactic	41.81±1.44	18.35±1.44
Tartaric	54.56±2.16	29.57±1.44
Malic	75.97±0.72	15.81±0.72

3 CONCLUSION

The present study shows that the functional properties of gelatines extracted from head of Mackerel and Blue Whiting were significantly ($p<0.05$) affected by the type of the organic acid used. Functionality tests indicated that the emulsifying and foaming properties of gelatines from Mackerel were significantly ($p<0.05$) higher than gelatines from Blue Whiting. These results show that gelatines from Mackerel head could be used as a functional ingredient for food and industrial applications.

References

1 M.C. Gómez-Guillén, M. Ihl, V. Bifani, A. Silva and P. Montero, *Food Hydrocolloid.*, 2007, **21**, 1133.
2 M.C. Gómez-Guillén and P. Montero, *J. Food Sci.*, 2001, **66**, 213.
3 B. Giménez, J. Turnay, M.A. Lizarbe, P.Montero and M.C. Gómez-Guillén, *Food Hydrocolloid.,* 2005, **19**, 941.
4 M.M. Bradford, *Anal. Biochem.*, 1976, **72**, 248.
5 J.F. Zayas, 'Solubility of proteins' in *Functionality of Proteins in Food*, Springer-Verlag, Berlin, 1997, p 6.
6 K.K. Agyare, K. Addo and Y. Xiong, *Food Hydrocolloid.*, 2008, **23**, 72.
7 Q.A. Fernandez and M.T. Macarulla, *Plant Food Hum. Nutr.*, 1997, **51**, 331.

Measuring Sustainability

LCA FOR CO-PRODUCT EXPLOITATION

Ulf Sonesson

SIK – The Swedish Institute for Food and Biotechnology, P.O Box 5401, SE 40229 Göteborg, Sweden, www.sik.se, telephone +46 (0)10 5166600, ulf.sonesson@sik.se

1 INTRODUCTION

When by-products or waste streams from food processing are being exploited into more valuable products instead of just being wasted, it will have environmental implications. These implications can be said to be twofold. One difference is about reduced need for waste management and the emissions caused by that, and the second effect is a result of the fact that the new products produced might replace existing products on the market, hence decreasing the environmental impact of producing these products. These complicated connections need to be considered when evaluating co-product exploitation, in the light of today's urgent need for environmental improvement from business. This article is a synopsis of a book chapter,[1] where the methodology and case studies are described in detail.

2 LIFE CYCLE ASSESSMENT (LCA)

Life cycle assessment is a methodology developed to quantify and evaluate the environmental impact of products and services. The method is described generally in,[2] and more specifically for applications in the food area in.[3] In principle an LCA follows a product from "cradle-to-grave" and quantifies all resources needed and all emissions caused to produce the product under study. For food products this means that production and transport of fertilisers and other inputs to agriculture is included, as well as energy used at the farm, in transportations, industrial processing, packaging, retail and in the households

2.1 LCA as a method to evaluate co-product exploitation

In the case of evaluating co-product exploitation, the need to perform full LCA's is limited. The interesting question is in first instance; "will we do better or worse for the environment by implementing the new process?". With that said, the focus should be on the changes incurred by the new process, i.e. it is a comparative study that is needed. The first step is to describe the present system, including waste management and what products are produced and in what amounts. Thereafter the new process is described, and by comparing these systems differences can be identified. The differences will often include present by-product use as e.g. feed, as well as direct energy differences in processing. Both systems need to deliver the same functions in order to be comparable. If system A delivers a function which is not present in system B, system B needs to be complemented with an "external production". An example of two systems using a co-product for different purposes is shown in Figure 1, as an example of what can be included in a study. The white boxes are those included in the original system, and the shaded boxes are external production included to fulfil the same function.

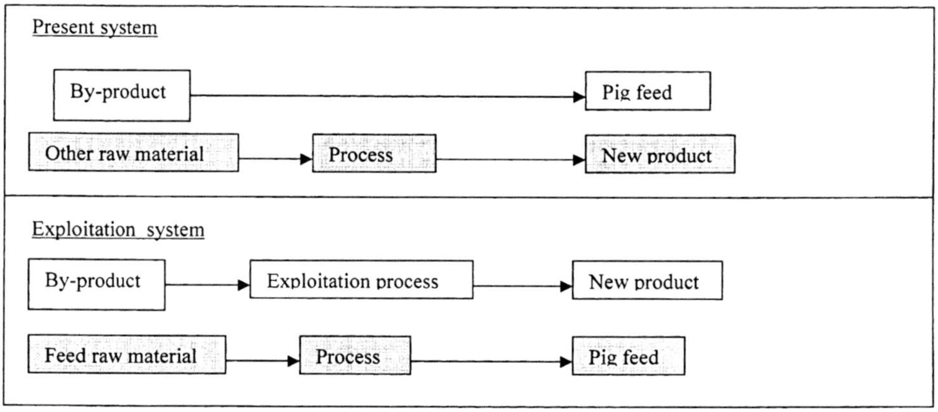

Figure 1 *Principal description of comparable systems of co-product exploitation*

3 CASE STUDIES – RESULTS

The case studies originate from the EU FP6 project REPRO (contract no. 006922). The first example is a rather simple and straightforward case, which facilitates an understanding of the principles, whereas the second case is more complicated, but gives an understanding of how a study is performed in practice.

3.1 Use of Brewer's Spent Grain in Extruded Snack Products

This is an example of a case where by-products are used to replace virgin material without altering the product. Hence the comparison includes for the present system the production of raw materials for the snack, the waste management for the by-product and the processing itself. For the up-grading system the same activities are included. There were also two alternative ways of handling the brewer's spent grain (BSG) besides the proposed up-grading; landfill or use as feed. Since these three scenarios deliver different products ("Landfill": no products, "Feed": animal nutrition, and "Snack": snacks) the results must be adjusted to make the scenarios comparable. This means that all scenarios should provide the same products, and if all products can not be produced from BSG, alternative means of production must be used, and the environmental impact and resource use for these alternative, or complementary, processes is added to the scenarios total (see Figure 1). In short, BSG replaced rice flour, other ingredients were the same.

In Figure 2 the results for contribution to global warming is presented. The BSG snack is the preferred option, and landfilling the worst, and use as feed intermediate. The reason is that in the "Feed scenario", alternative production of ingredients causes more global warming gasses than the alternative feed production in "Snack scenario". The "Landfill scenario" is charged with both alternative ingredients and alternative feed production, and in addition the landfilled BSG cause emissions of the green house gas methane.

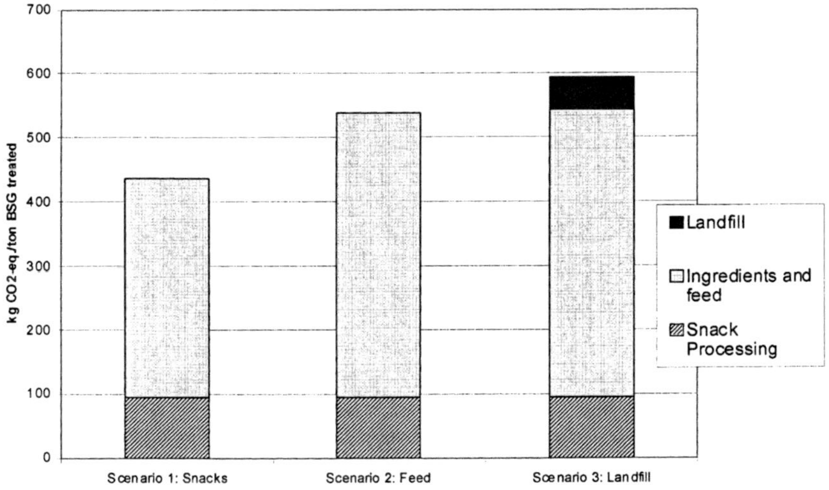

Figure 2 *Results for global warming potential, for the comparison of different ways of using/disposing BSG*

Generally the results show that it is beneficial for the environment to use BSG as a snack ingredient instead of using it as feed, and even more beneficial than deposit the BSG on landfills. The reason is the fact that the "BSG comes for free" as compared to the alternative ingredients, and it is environmentally more efficient to produce the feed directly than to use the BSG as feed and produce alternative ingredients (rice meal).

An interesting observation is that if the assessment would have been performed without the systems perspective and hence not included the decreased use of raw materials and the changed feed production from BSG to grain (grey bars in Figure 2), the conclusions would have been different. Using BSG for snack production would have performed better than landfilling it, but not better than using it for feed. Thus, the systems perspective gives a deeper analysis and understanding of the environmental effects of up-grading of by-products.

3.2 Using Enzymes to Up-grade Red Cabbage Trimmings into Valuable Products

When cabbage is processed, a certain percentage is lost as trimming, which in the case study is used as cattle feed or composted. By using enzymes, red cabbage trimmings are partly degraded and valuable components can be extracted. The products achieved are pectin, MHR ("modified hairy regions", a health promoting substance), dietary fibre and a phytochemical rich cabbage juice. The aim of this study was to answer the question "Among the different options to deal with red cabbage trimmings (RCT), which is the alternative with least environmental impact associated?" so the function of the system is to treat this specific food by-product. The basis for comparison was 1,000 kg of RCT.

Included activities are the by-product management or the composting in the present system and the up-grading process for the novel system. In addition the "additional production" is included depending on what is produced in the scenario.
In order to make the scenarios comparable, their function needs to be equivalent, that is to say, all the scenarios need to be defined so that they include the production of the same/equivalent products. We choose to perform a so called "systems expansion", where if not all products can be obtained from the RCT, alternative means of production must be used, and the environmental impact and resource use for these alternative, or complementary, processes is added to the scenarios total (see Figure 1). In summary, the scenarios defined for comparison in this study are:

Scenario 1: RCT for value-added products, Alternative production of cattle feed (silage), Production of compost from organic waste

Scenario 2: RCT for cattle feed, Alternative production of dietary fibre, juice, pectin and HMR, Production of compost from organic waste

Scenario 3: RCT for compost, Alternative production of dietary fibre, juice, pectin and HMR, Alternative production of cattle feed (silage)

In order to make the scenarios equivalent, alternative means of production for the products obtained by the model also need to be inventoried: dietary fibre, phytochemical-rich juice, pectin and MHR.

In Figure 3 the global warming results for three scenarios producing pectin, feed and compost are presented. The liquefaction performs better than the, the main reason being the fact that the alternative production of vegetable juice (carrot) is more environmentally demanding than the juice made by RCT. Alternative pectin production also plays a role, but smaller. This is explained simply by the fact that more juice than pectin is produced. It should also be noted that alternative production of MHR is not included in the results for the "Cattle feed" and "Compost" alternatives, thus the results for those alternatives are underestimated, but it is not possible to state how much as a result of data gaps.

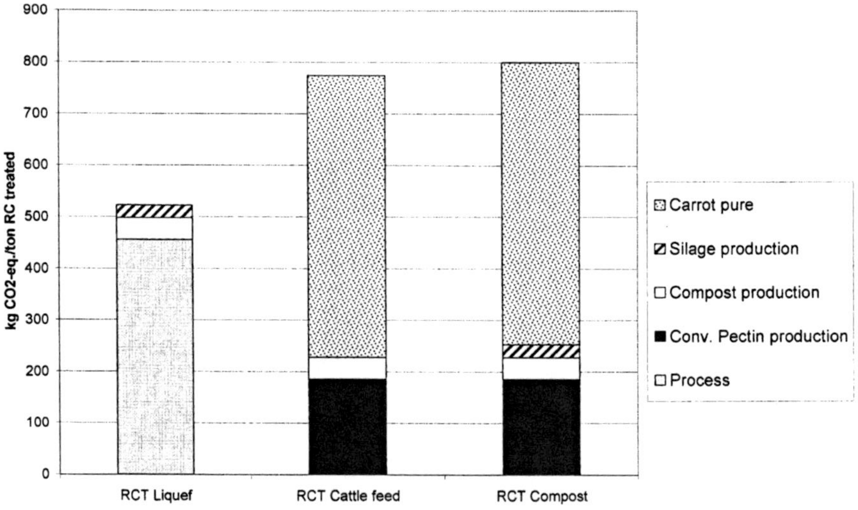

Figure 3 *Potential global warming effect for liquefaction of Red cabbage trimmings with maximum pectin production, compared to alternative routes.*

This case study clearly shows the importance of the avoided production of the products produced, i.e. the saved environmental impact due to a better utilisation of the raw material red cabbage. It also shows that it is important to understand that if by-products are to be up-graded instead of being used as feed the feed has to be supplied from another source. It is a trade-off between saved- and increased alternative production.

In both cases the production of cabbage juice is very important for the results, resulting from the fact that the volume of juice is large, hence the other scenarios are responsible for much alternative production. This reinforces the conclusion that a systems perspective, where all changes incurred by the new up-grading process must be included to get a relevant basis for decision.

4 CONCLUSIONS

In order to make environmentally sound decisions when developing new exploitation processes for food industry co-products, a holistic approach to evaluate the environmental effect is crucial. By using a method based in LCA this can be achieved.

References

1 U. Sonesson, Application of life cycle assessment (LCA) in reducing waste and developing co-products in food processing, In: *Handbook of waste management and co-product recovery in food production*, ed. K.W. Waldron, Woodhead Publishing Limited, Oxford, UK, 2009.

2 H. Baumann and A-M. Tillman, *The Hitch Hiker's Guide to LCA – an orientation in life cycle assessment methodology and application*, Studentlitteratur. Lund, Sweden, 2004.

3 J. Berlin, Life Cycle Assessment (LCA): An Introduction, In: Mattsson, B. & Sonesson, U. (eds), *Environmentally Friendly Food Processing*, Woodhead Publishing Limited, Cambridge, UK, 2003.

Integrated Approaches - Process and Chain

PROCESS OPTIMISATION

David Napper,[1] Jin-Kuk Kim[2] and Igor Bulatov[2]

[1]Euroteknik Limited, Bosworth Hall, Nr. Lutterworth, Leics, LE17 6LZ, UK
Email:dn@cseparator.com
[2]Centre for Process Integration, The University of Manchester, M60 1QD, UK

1 INTRODUCTION

There have been continual engineering developments in processing for various industries such as the "pinch process" for water and energy in the chemical and petroleum sectors. Since the 90's the Process Integrated Environmental Protection (PIEP) has had much use in the chemical industries to minimize the continual creation of substances harmful to the environment. This "Process Optimization" presentation is a paradigm of an effort to systemize the mentioned disciplines and add chemical plus by-product pinch, and look at how a holistic picture can be created to define the total impact including the CO_2 profile and carbon trading considerations in the food and beverage sector.

2 HOLISTIC APPROACH MANAGEMENT CONCEPT

The broad knowledge management concept consists the following 10 categories of intervention.

1. Dissemination model – to define, track and keep all stakeholders informed.
2. Interrelationships are defined and observed to cross feed particularly relevant information especially when sub system decisions need to be taken.
3. Process. At the heart of the entire exercise is the following of the production of the factory. This section includes traceability with input from documents, RFID technology and points of validation guided by standardized requirement for operator checkpoint affirmation, rejection or deviation from defined parameters and all inputs go into a third party database without the chance for corruption.
4. Risk analysis starting with HACCP model for the food factories, the problems they have encountered and the solutions used. The database of information will be examined with a case based reasoning (CBR) model will be used to help create solution models based on the aggregated information. The models will contain preventative and predictive measures that can be implemented.
5. People involved - to track in order to document what each actor has done, when they did it, and if they do all they are responsible for doing.
6. Process points / analysis with reading of data manually and automatically.

7. Collection of data allows for the transparent benchmarking of factories the prediction of what can be done to increase performance.
8. Data collection, find out what common needs are for the near 200 food factories involved.
9. Educational process was designed to create awareness and competence.
10. Finally an aggregate life cycle analysis LCA for the overall process chain.

3 PROCESS OPTIMISATION STRATEGIES

Optimizing the process and associated design study aims to gain in-depth understanding of production and manufacturing, and hence, improve the system performance (e.g. cost, efficiency, throughput, emissions, etc). Heuristics and engineering experience may be useful by supporting the decision-making which attempts to identify optimal or much improved processes (and/or its performance). However, such a non-systematic judgment often misses potentials or novel options for achieving better system performance. More importantly, without appropriate optimization study, all the degrees of freedom and their design interactions are not fully investigated, and this leads to very minor advances from current design or process. Then, how do we optimize the process? There are two trends in process industries.

- Mathematical programming with application of optimization techniques: System performance and operating characteristics (e.g. mass and heat balances, constraints, etc) are represented with mathematical format. Improved solutions are then sought within a solution space, by using deterministic or stochastic optimization techniques. The decision of best design structure, together with determining optimal operating parameters can be made.
- Graphic-based process design methods: System information is mapped to key design variables and these variables are graphically manipulated, with which optimal conditions (targets) are obtained. Targeting information is then used to provide strategies and guidelines for the design. For example, water minimization study, two key parameters, concentration and mass load, are selected, and overall performance of water-using systems are graphically represented by "water composite curve". Minimum water consumption is targeted from the composite curve, from which target information is then used to complete the network of water-using systems.

A holistic approach is envisaged by examining integrated features with these design methods, which is essential to ensure high sustainability and economics of the process, as well as to enhance current engineering, operation and design practices.

Improving energy and water efficiency in food industry is one of key to ensure cost-effective manufacturing, leading to strong competitiveness of the company in the market. Water savings can be achieved through strategic implementation of water reuse between water-using operations. Further minimisation on freshwater usage can be obtained by regenerating water which is then and recycled. Improving water efficiency in the process not only contributes reduction of operating costs, but also, more significantly, provides considerable incentives to the design and operation of effluent treatment, as significantly less volume of wastewater is to be treated, which reduce capacity of effluent plants and its associated operating costs. Design methodology developed by Wang and his co-workers[1,2,3] has been proven to be very effective in process industries, because the method systematically provides realistic minimum water requirement for the site, as well as conceptual design guidelines for the debottlenecking of water systems.

Improving energy recovery has gained recently considerable attentions from food industry in order to respond climate change and improve sustainability of the company's operation. As various design options and mechanism are available for generating, distributing and consuming energy in food industry, it is not straightforward to determine most effective design for energy systems and operating practices of utilising energy. Heat Integration (or Pinch Analysis) technique has a successful track record in improving energy efficiency and heat recovery, and various graphical tools (e.g. hot and cold composite curves, grand composite curve, total site profile, etc) are available for targeting and optimising site-wide energy infrastructure.[4,5]

In food industry, water use is closely linked and interacted with energy systems. For example, heating is required for the water used in sterilisation and process-washing. Water is heated up during the operating, is discharged at high temperature. Water used for cooling or condensate return in steam systems contains valuable energy. Water minimisation study cannot be separated with energy saving study, and vice versa. This is especially true for food industry, as considerable amount of heat is often added to water streams and considerable amount of waste heat is discharged without full recovery.

The schematic diagram is shown in Figure 1, which explains basic concept and the importance of simultaneous consideration of water and energy systems. In Figure 1(a), freshwater is supplied to two water-using operations and water is then discharged with parallel arrangement. Necessary heating or cooling through, typically, heat exchanger, is provided according to process requirements. This conventional practice can be significantly improved by implementing water reuse and heat recovery, as shown in Figure 1(b). Water reuse between operations is to minimise water consumption, while heat recovery between streams are envisaged, which can avoid unnecessary utilities (e.g. steam, cooling water).

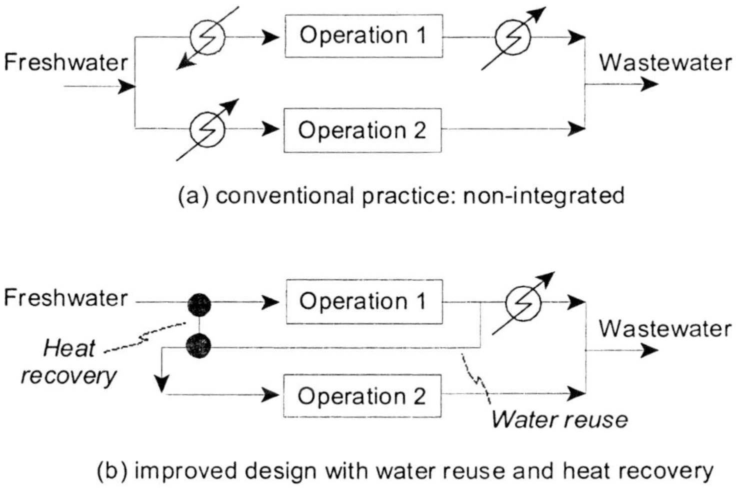

Figure 1 *Simultaneous energy and water minimisation*

However, complexities in finding best energy recovery options as well as water reuse schemes "together" are very high, as there are strong design interactions between water and energy systems. Recently, Kim and his co-workers developed design methods which carries out simultaneous analysis of energy recovery in water systems, using either graphic-based method[6,7] or mathematical optimisation method.[8] Both water-pinch and energy-pinch concept has been accommodated in the design frameworks, however, a step change in the methodological procedure has been made to consider simultaneous interactions between water reuse and energy recovery.

Design procedure follows two-step approach: Targeting and design. In targeting, theoretical minimum requirements for freshwater and utilities (hot and cold) are obtained by the aid of graphical manipulation of streams data (i.e. water flowrate, contaminant levels, temperature). Design of water and heat recovery networks is carried out to achieve the "target", using Two Dimensional Grid Diagram, which exploits potential network arrangement of water streams, subject to energy recovery constraints. An industrial-size case study conducted recently[8] showed that overall 18 % improvement in annualised cost, from the integrated approach (compared to the base case in which only water minimisation is performed). This case study clearly demonstrated why a holistic approach for water and energy systems is required, and significant benefits can be gained by using simultaneous integration method.

4 CONCLUSIONS

Recent research strongly indicated that in the fish industries water and energy are wasted, that effluents and waste treatment are not optimized and therefore not sustainably managed. Too often, effluents and by-products are being treated as waste and discharged for example such as ensilage in landfills. This creates unsustainable consequences to the environment. Significant methodological development on pinch analysis for water and energy saving have been made and its successful industrial applications have been observed. A holistic approach has been taken to identify most appropriate actions for improving system performance and efficiency, and to provide cost-effective upgrading strategies and practical engineering solutions. Also, the design procedure allowed engineers to understand what operational or structural changes should be systematically made, in order to spare the environment and improve sustainability. In addition, economic considerations were taken into account as well as process improvements affecting the overall value chain and public acceptance/support. In Europe, many organizations such as the RUBIN foundation in Norway and the SEApro network in France estimates that fish industries waste values could be increased by at least five fold if valorisation of by-products was performed. Part of this presentation will also evaluate the value chain to pin point where the most obvious interventions could be initiated to add value to the materials being discarded and propose solutions for further processing.

References

1 Y.P. Wang and R. Smith, *Chem. Eng. Sci.*, 1994, **49**, 981.
2 W.J. Kuo and R. Smith, *Chem. Eng. Sci.*, 1997, **52**, 4273.
3 M. Gunaratnam, A. Alva-Argáez, A. Kokossis, J. Kim and R. Smith, *Ind. Eng. Chem. Res.*, 2005, **44**, 588.
4 J. Kim and R. Smith, in *Encyclopedia of Chemical Processing*, Marcel Dekker, 2005, p2165.
5 R. Smith, Chemical Process Design and Integration, Wiley, UK, 2005.

6 L. Savulescu, J. Kim and R. Smith, *Chem. Eng. Sci.*, 2005, **60**, 3279.
7 L. Savulescu, J. Kim and R. Smith, *Chem. Eng. Sci.*, 2005, **60**, 3291.
8 B. Leewongtanawit and J. Kim, *Chem. Eng. Proc.*, 2008, **47**, 670.

LAST MINUTE MARKET. INCREASING THE ECONOMIC, SOCIAL AND ENVIRONMENTAL VALUE OF UNSOLD PRODUCTS IN THE FOOD CHAIN.

A. Segrè, L. Falasconi and E. Morganti

Department of Economy and Engineering Agricultures, Alma Mater Studiorum – University of Bologna, Viale G. Fanin 50, Italy

1 LAST MINUTE MARKET, A LOCAL DEVELOPMENT PROJECT

1.1 The Last Minute Market model

For different reasons, the developed economies produce a growing quantity of food surplus, in every single phase of the food chain, starting from the agricultural production and ending with the retail system. Basically, this surplus is made up of products not saleable for several reasons (close to the expiry data; defects/damages in packaging, etc..), but still safe to eat (Jones[3-4-5]).

The potential supply is provided by the so-called profit organizations, i.e. firms, processing industries, food shops, retail stores and the like. For these companies surplus represents actually a cost (transport and destruction of the products), while for society as a whole, a negative externality (environmental pollution). In this picture the part of the (potential) demand is played by a number of charity associations or non profit organizations that assist directly and indirectly poor and marginal people (Birch[1]).

Linking surplus (supply) and deficit (demand) could counterbalance this "imbalanced" food market: a way to provide such a linkage has been developed by Last Minute Market (LMM), the University of Bologna spin-off company, which has defined procedures related to sanitary-health issues, tax-related issues, logistics, nutritional issues and marketing in order to transfer in safety the products (Segrè[7]).

The main objective is to create a contact between supply (for profit organizations) and demand (non profit organizations): the scheme in which the exchange is performed works throughout as a gift transfer between the profit and non profit organizations.

This creates a local, dynamic and stable network between companies, agencies and associations, formed by solid interactions and the exchange of goods and assets.

The advantages that it generates are economic (since the management of excesses creates costs for companies both in terms of internal management and waste removal), environmental (since waste removal and its transportation create pollution), social-health care (since food products are recovered and sent to people in need as much as possible) and educational (since it raises public awareness on the problems of food waste) (Falasconi[2]).

1.2 Objectives and Methods

This project was born as research in the Department of Agricultural Economics and Engineering in 1999, in order to analyse surpluses in agro-food production chain.

The study carried on the exploratory phase, the experimental phase and the consolidation and expansion phase with the aim to collect information and data causes of the phenomenon in-depth

The objective of providing reliable results, having as their reference point the values of the experimental activity (available pattern) for a mid-size supermarket and a group of benchmark school cafeterias. A fundamental aspect of this analysis, which should be underlined, was that that hypotheses used always aimed at undervaluing rather than overestimating the amounts obtained, given that the estimate was strongly dependant on the available sample. To obtain these results, single variable statistical techniques were used, like the Box-Jenkins method (Segrè[6]).

The second objective of our analysis was focused on determining the amount of calories available from the benchmark supermarket based on the types of products recovered and projecting that data on the national level. In addition to the environmental impact, the social impact was estimated, leading to the determination of a number of subjects who could be completely aided on the micro level (benchmark supermarket), referring to the daily average calories needed per individual based on the data from the FAO and LARN sources.

To calculate the calories available in the products recovered per individual department or per individual month, given the data available, an index of the daily calories needed was used based on the amounts recovered each month and the department analyzed, calculating its correlation. A (direct and linear) relationship between the amount of products collected (instrumental variable) and the average monthly calorie index per department (dependent variable) was estimated, which, appropriately multiplied, allowed us to obtain the exact number of calories produced. The overall model was estimated using the OLS method, using several dummy variables to consider the importance of department-specific factors.

2 CASE STUDIES

2.1 1st Case Study: Supermarket Recovery

The new logistic management of unsold products (storage and distribution) that leads to their recovery, keeping, as much as possible, their original destination for which they were produced intact (human consumption), can mean that an average size supermarket (6,000-7,000 m^2) recovers every years about 65 tons (net weight) of food products. From an organoleptic point of view, these products are completely identical to the products on the shelves in the sales area, but they are not sold for aesthetic reasons only (external product packaging damaged, product near its expiration date but not expired, damaged produce, etc).

Here we report the example of a supermarket where Minute Market started working in 2005, the Leclerc-Conad supermarket in Modena with a surface area of 6,300m^2. The trend over the last four years is listed in Table 1.

By studying the type of product that can be recovered in detail, a fundamental aspect emerges: this new logistic-management method for unsold products allows us to recover and distribute fresh and very fresh products, which, until now, has encountered numerous difficulties.

Table 2 shows the total amount of annual products recovered, the majority (65.37%) is fruits and vegetables. The second area that is significant in terms of weight is the meat department, that represents more than 18%. Next, we have Dry Goods (cans, pasta, cookies, snacks, fruit juice, flour, oil, etc) that makes up 10%. Even if the amount and thus the percentages (4.76%) are less, the Dairy sector is important because it can be widely distributed and given the product's nutritional value. Finally, we have the Bakery, where we recover 949 Kg, or 1.46%, of products (See Table 2).

Table 1 *Unsold Product Recovery Trend (Amount and Economic Value) at the Leclerc-Conad (MO) Supermarket.*

Year	*Kg*	*€*
2005	92.4	251,466
2006	61	161,339
2007	49.9	203,613
2008	53	197,828
Total	*256.3*	*814,246*

Table 2 *Percentage the individual departments represent out of the total of products recovered and distributed.*

	Dry Goods	*Dairy*	*Meat*	*Produce*	*Bakery*	*Total*
Weight (Kg)	6,500.0	3,094.0	11,966.5	42,490.5	949.0	65,000
Percentage	10.00%	4.76%	18.41%	65.37%	1.46%	100%

It must also be considered that for health and hygiene related reasons as well as for mere practical conditions (for example, boxes of pasta, flour, open cookies, left over leafs of lettuce, etc) that not all the products recovered can be used for human consumption so we decided to try to attain another goal, what we call "second best," i.e. animal consumption. This target is only considered secondary due exclusively to the fact that the main goal is to keep the original destination of the food products, i.e. human consumption.

Based on these two goals, 96.5% of the products recovered were able to be used for human consumption while the remaining 3.5% *was considered suitable for animal consumption.*

2.2 2nd Case Study: Recovery from Dining Halls

The second case instead deals with the new logistical management of meals unserved (collection and distribution) by company and public and private school cafeterias. In this case, given that the collection occurs immediately after the cafeteria is closed to the public, all the food recovered is edible as per the law and suitable for human consumption. The follow refers to one of the projects started in Italy and, in detail, involves a group of 8 elementary and middle schools in the City of Verona for academic year 2007-2008. With daily food recovery, which differs from school to school, it was possible to save 10,166.24 Kg of ready to eat meals from destruction, resulting in about 20,000 meals in all.

The extremely innovative aspect of this activity is the fact that by collecting ready to eat food, it only needs to be served since it was already cooked and heated when it was picked up. If, on one hand, this aspect provides significant advantages, on the other, it significantly complicates the recovery activities since these foods can spoil easily, require packaging, are extremely complicated to transport and consume. At least in the very beginning, these aspects make this type of recovery extremely complicated.

Table 3 *Recovery Results from 7 School Cafeterias.*

TYPE	*AMOUNT (Kg)*	*PERCENTAGE*
Pasta	4,384.70	43.13%
Main Course	1,687.60	16.60%
Side Dishes	2,001.73	19.69%
Fruit/Yogurt	435.12	4.28%
Bread	1,511.72	14.87%
Miscellaneous	145.38	1.43%
TOTAL	10,166.24	100.00%

It is important to underline that with this type of recovery and the constant and punctual monitoring of unserved foods, the City of Verona received important information on consumption trends that allows it to fine tune the amount of food prepared each day while still respecting the guaranteed portions for each child.

During the reference period, it emerged that, on average, 18.1% of the meals prepared were recovered every day. This excess is not due to the poor management of the meal preparation department but rather to the fact that the portion size of the individual means is set forth by sample tables prepared by nutritionists based on the standards which do not always correspond to the tastes and appetites of the children. Second, slightly more meals must be prepared than the base number in order to provide seconds to whoever asks for them.

3 EVALUATION OF THE IMPACTS FROM THE USE OF THE RECOVERED PRODUCTS

3.1 Economic Impact

A part of the environmental benefit from the reduction of the about of products disposed of directly translated into a reduction of disposal costs for the supermarket. The entire environmental impact for the company serving as a donor can vary widely due to various internal and external factors for the company, which impact the local waste management cost.

The Italian tax laws also impact the economic management of the unsold product by the sales point, which encourages the donation of these products to local charities. Based on the type of product, there are produces for VAT reimbursements and specific instructions for the product management.

Among the factors that impact the decisions of companies implementing recovery plans are also influential political trends and their application through European and domestic directives and regulations. Today, local governments continue to apply different disposal tariffs in accordance with the "polluter pays" principle. It is important to include the possibility of enjoying a discounted disposal tariff rate in estimating the company's economic benefit if set forth by the current law, a sort of incentive to behaving virtuously to reduce and prevent waste.

Among the effects that produce an economic return for the company there are: the waste management and monitoring service (managerial instrument), and media visibility in the local context.

The beneficiaries enjoy an economic advantage similar to the commercial value of the goods recovered and the savings obtained free up resources to invest in order to improve the quality and quantity of aid services offered.

3.2 Nutritional Impact

Based on the amount of food products recovered and distributed by type of supermarket and the related calories for each type of product (defined specifically and in-detail by LARN), the annual amount of calories available can be quantified, which total 149,650,994.17 Kcal (See Table 4).

Table 4 *Total calories produced and distributed in a year per mid-to-large size supermarket.*

	Dry Goods	*Dairy*	*Meat*	*Produce*	*Bakery*	*TOTAL*
Kcal	39,107,977	16,309,602	46,422,354	34,674,687	13,136,372	149,650,994

To complete this data, which might not be important, the number of people who could annually benefit from these products was estimated. In this case, we need to highlight that only 70% of these calories are intended for human consumption. Considering that the average daily consumption per person is 3,680 Kcal (FAO data per an average period with an average body size participating in moderate activity), 28,466 people can be helped per year and about 71,000 meals provided (breakfast, lunch, dinner) a year per supermarket, which are 190 meals a day.

The analysis can be conducted by following a second path, dictated by nutritional needs and verified by the empirical daily nature of the daily distribution. In fact, it can be considered that each individual person consumes about 0.5 Kg of food per meal a day, which means that 70% of the 146 tons of products recovered by the individual supermarket (the part for human consumption) can provide 204,000 meals a year, or 560 meals a day.

Table 5 *Estimates of the total calories that can be produced and distributed per year from the total number of domestic supermarkets (in billions).*

	Dry Goods	*Dairy*	*Meat*	*Produce*	*Bakery*	*TOTAL*
Kcal	15.40	6.03	17,10	12.87	3.27	54.68

Reporting and simplifying everything to the number of potential people who could enjoy the products and meals potentially provided, we see 10,402,216 aided a year (considering the only 70% of the calories is for human consumption and an individual consumes 3,680 kCal a day per year). This is a total of 26,000,000 meals a year.

If we take the second path, i.e. the amount of food that can be consumed per meal (0.5 Kg pro capita), we get 70,000,000 meals a year from all the Italian supermarkets.

3.3 The LMM replicability

The LMM model has been modified in order to manage the surplus of different kind of products, not only food. Below, we will briefly illustrate the variants that the Last Minute Market project had implemented.

3.3.1 Last Minute Market – Harvest. In agriculture, the phenomenon of hyper-production exists and particularly involves the produce sector. The phenomenon of not harvesting "fruits" occurs since they either do not meet the required quality standards or because they find the market demand is less than the supply. These crops are therefore destined to literally rot in the field, constituting an unacceptable waste of food with high nutritional value. LMM - harvest re-introduce the ancient method of gleaning, which etymologically means "collecting ears of grain from the already harvested fields or collecting what remains", involving in the action the charity operators and, if possible, people receiving aid.

3.3.2 Last Minute Market – Seeds. Seeds whose germination rate is slightly lower than the minimum limit set forth by the E.U. regulations cannot be sold on the market. There only use is incineration since this material is potentially at great risk of biological pollution. The goal is to avoid this waste by recovering the seeds and providing them, free of charge, to farmers in developing countries. In this case, the question of proximity is set aside. The end goal is not only recovering the goods but also providing suitable instruments to the poor farmers in these countries, trying as much as possible to respect their habits and agricultural traditions.

3.3.3 Last Minute Market – Pharmacy. The unsold drugs are taken off the shelves 1-6 months before their expiration date. In this case, deals with conserving and administering the drugs recovered, which must naturally be done by authorized medical personnel.

3.3.4 Last Minute Market – Book. Unless food products, books do not "expire" but, after a certain period, from a few months to a year (up to 8), they are taken off the shelves because it's thought that readers no longer want them. Instead of ending up in the incinerator and being disposed of like food, they are ground up. The editor manages to make a small profit by selling the paper, which is cut into by the costs incurred to transport them to the recycling centers.

However given a potential book offer, a corresponding potential demand can be found in the public institutions.

References

1 G.G. Birch, K.J. Parker, J.T. Worgan, *Food from waste*, Applied Science Publishers, London, 1976.

2 L. Falasconi, A. Segrè, *Abbondanza e scarsità nelle economie sviluppate. Per una valorizzazione sostenibile dei prodotti alimentari invenduti*, Franco Angeli, Milano, 2002.

3 T.W. Jones, *Analyzing retail food loss,* biocycle – Journal of composting & organics recycling Vol. 46, n. 12, Emmaus – Pennsylvania, 2005, pp. 40.

4 T.W. Jones, *Food loss and the American household*, biocycle – Journal of composting & organics recycling Vol. 47, n. 3, Emmaus – Pennsylvania, 2006, p. 28.

5 T.W. Jones, *Food loss on the farm*, biocycle – Journal of composting & organics recycling Vol. 46, n. 9, Emmaus – Pennsylvania, 2005, pp. 44.

6 Segrè, *Elogio dello –SPR+ECO. Formule per una società sufficiente*, audio-video-libro, Emi, Bologna, 2008.

7 A. Segrè, *Lo spreco utile. Il libro del cibo solidale. Trasformare lo spreco in risorsa con i Last Minute Food e Book*, Pendragon, Bologna, 2005.

Energy Recovery and Technologies for Water Recovery and Recycling

BIOMETHANE AND BIOHYDROGEN FROM FOOD BYPRODUCT

C.L. Hansen, [1] J.S. Dustin[2] and R.S. Thompson[3]

[1]Professor, Departments of Nutrition, Dietetics and Food Science and Biological and Irrigation Engineering, Utah State University
[2]Ph.D Candidate, Department of Biological and Irrigation Engineering, Utah State University
[3]Technician, Utah State University

1 INTRODUCTION

1.1 Induced blanket anaerobic digester

Often, the major purpose of anaerobic digestion in wastewater treatment is to reduce waste sludge volume and activity by fermenting the waste in the absence of oxygen. A major advantage of anaerobic digestion is the relatively low amount of sludge generated from destruction of organic matter which is due to the slow growth of anaerobic bacteria. The relatively long doubling times of anaerobes require either long residence times (20-40 d) to allow populations to accumulate to levels permitting significant treatment or separation of hydraulic and solids residence times to permit high-rate treatment. Methods for accomplishing this include separating the active solids from the effluent and recycling them to the digester, by directly utilizing an attached growth process, or through the use of an induced bed or blanket of solids that self-select to remain in the reactor as the substrate passes through.[1-3] These processes generally require a relatively low suspended solids concentration in the influent to avoid dilution of the active biomass and to avoid operational problems in the digester (plugging, sedimentation, short circuiting, etc). This reliance on differentiation of solids and substrate poses a problem in treatment of some wastes, particularly those with high concentrations of suspended solids like those associated with food processing wastes and livestock manures.

The induced blanket reactor (IBR) anaerobic digester (Figure 1) was developed at Utah State University in the early 1990's by Dr. Conly Hansen to address the operational limitations of applying high rate treatment techniques to high strength, high solids substrates. It relies on biomass retention in the digester to provide an accumulation of active solids, similar to the operational principles first proposed by Dr. Gatze Lettinga for his UASB reactor design.[1] UASBs operate on the principle of separation of solids and hydraulic retention times (SRT and HRT) via granulated biomass retention. The active biomass acts as an effective filter, metabolizing available carbon to CO_2 and CH_4. In the UASB, influent enters the reactor from the bottom via a diffuser to provide equal distribution through the sludge blanket and minimize short-circuiting. Substrate transits the sludge blanket vertically, converting chemical oxygen demand (COD) to biogas. Biogas is directed via the deflector baffles to the 3-phase separator, while treated liquid effluent flows around the baffles and the separator and over a weir to the reactor outlet.

The Induced Blanket Reactor (IBR) is an upflow design that preserves some of the biomass concentrating characteristics of the UASB while allowing for the treatment of complex high solids and wastewaters such as dairy manures. The septum baffle separates the mixed liquid volume from the headspace, while the diffuser plate enhances substrate diffusion into the sludge blanket and the auger keeps the gas outlet in the septum baffle clear of solids. Liquid effluent exits the reactor via an inverted trap above the septum.

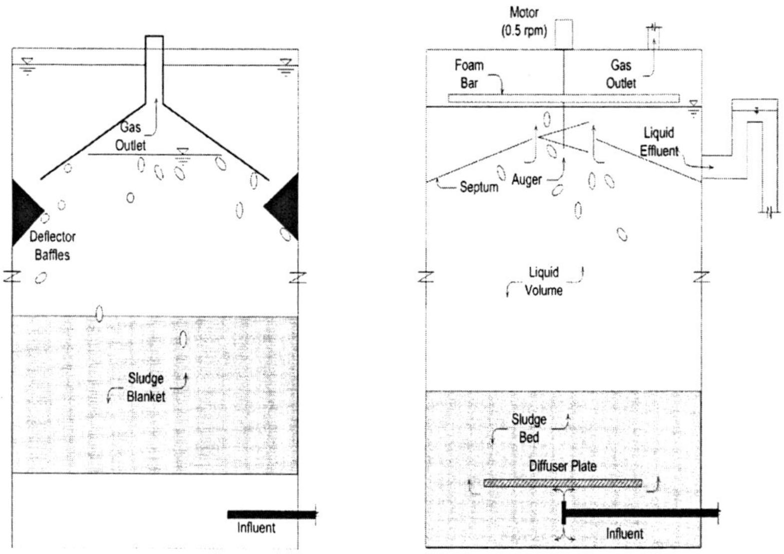

Figure 1 *Cutaway views of generalized Upflow Anaerobic Sludge Blanket (UASB) (L) and Induced Blanket (IBR) (R) reactors.*

The differences between the two reactor types are derived from the need to address the high solids content of the influent. Inlet and outlet conditions and the gas separation mechanism were modified to prevent plugging in the reactor. In addition, the HRT for the IBR is several times that required for the UASB. The principal advantages of the IBR system are that the system a) reduces the need for solids separation over what would be required for other high-rate systems, b) permits the processing of all the material in the feed stream, and c) with a 4-day HRT requires 5x less volume than a similarly performing plug or complete mix digester with a 20-d HRT. The IBR has been successfully implemented with over 4,000 m^3 of installed capacity in the US and Canada.

In order to model any reactor, an understanding of reaction kinetics must be coupled with an adequate model of the mixing and contacting patterns in the system.[4] Broadly speaking, there are two extremes of mixing that may occur, one of which will generally be ideal for any reaction.[5] In an ideal completely stirred tank reactor (CSTR), the contents are completely mixed such that the properties are uniform throughout the reactor. Mathematically, this can be represented by an exponential decay function (Equation 1).

Plug flow is the condition where elements of a fluid entering a vessel at the same time move through it with constant and equal velocity on parallel paths and leave at the same moment. Residence time distribution (RTD) studies in which an appropriate tracer is injected into the system then monitored as it transits the reactor are generally used to determine RTD.

$$C_t = 1 - e^{-t} \tag{1}$$

C_t = Tracer concentration at time t

t = Sample time

Lab and full scale RTD studies in the IBR have shown that the most appropriate model for the IBR appears to be a CSTR. Although random elements of short circuiting and dead space were observed for individual reactors and studies, over a series of nine treatments at varying temperatures and using different tracers, CSTR behavior was dominant.[6] The lack of consistency in these variations between studies coupled with the very good fit of the combined data to an ideal CSTR model leads to the conclusion that the IBR is can be represented symbolically as a CSTR, but that the lack of mechanical mixing may impose a non-homogeneity of reactor contents and flow patterns such that while the overall effect is that of a CSTR, the mixing is not perfect and the reactor is subject to some arbitrary flow. In light of these data, it is recommended that designers model the IBR as a real CSTR (Figure 2) that includes components of a stirred tank reactor, a bypassing flow, and a dead zone where no mixing occurs.

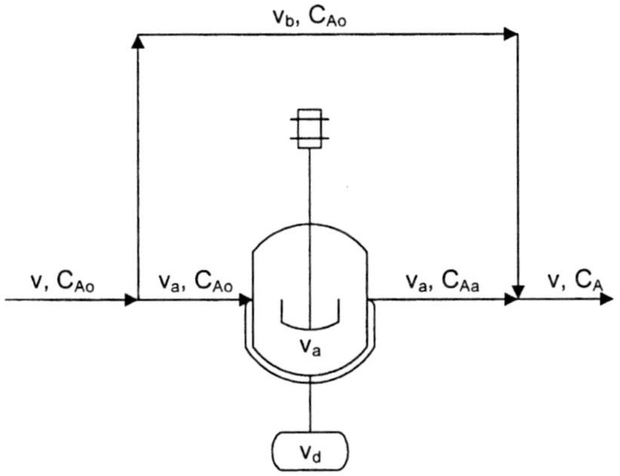

Figure 2 *Real CSTR*

It should be noted that the mixing parameters derived by work to date are necessarily limited to the behavior of the liquid fraction of the reactor contents. By definition, as a retained biomass reactor, the IBR maintains a separation between the solids retention time (SRT) and hydraulic (liquid) retention time (HRT) in the reactor. This requires that mixing energy be maintained below the threshold that would provide for uniform distribution of solids in the reactor and eliminate the difference between SRT and HRT.

1.2 Biologically produced hydrogen

The global power supply is currently 84.8% fossil energy.[7] World energy consumption is expected to climb steadily over the next thirty years as a result of economic growth, particularly in developing nations and population growth throughout the world. In the search for an alternative fuel to supply this energy demand, special consideration has been put on a fuel that not only supplies the world's energy demands, but is also a cleaner option to the fossil fuels used today. One source of energy which has received special attention for meeting these requirements is hydrogen fuel. Hydrogen gas is termed as a clean fuel because water instead of greenhouse gases is produced when combusted. It has an energy yield of 122 kJ/g, more than twice that of hydrocarbon fuels.[8] Hydrogen has the potential to lessen the worlds dependency on fossil fuels, but further research and technology is needed before a sustainable hydrogen economy can be established.

Despite its clean and green nature, most hydrogen is currently produced from fossil fuels, such as natural gas, oil, and coal as seen from Figure 3. An alternative hydrogen production source is anaerobic fermentation. Anaerobic fermentation produces hydrogen that is less energy intensive than chemical or electrochemical methods because it is normally carried out at ambient temperature and pressure.[9] Anaerobic production of hydrogen is an exciting new area of technology development that offers the potential to produce usable hydrogen from a variety of renewable resources such as organic wastes (*e.g.* food processing waste and animal waste).[10]

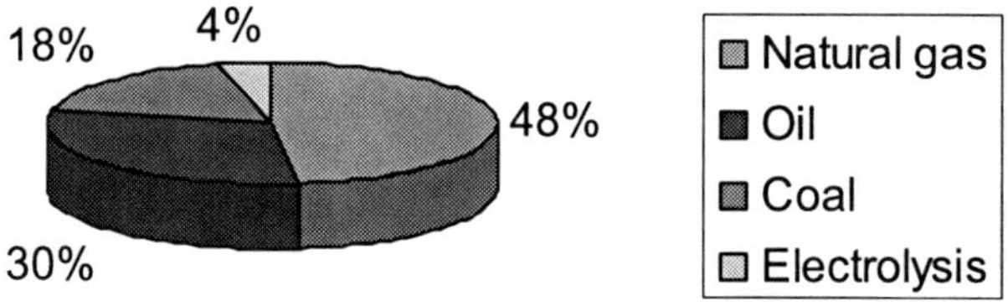

Figure 3 *Overall world wide hydrogen production sources by percent.*[11]

Current research has studied different types of substrates for biological production of hydrogen. The major criteria for substrate selection are availability, cost, carbohydrate content, and biodegradability.[12] Commercially produced food products, such as corn and sugar, are not economical for hydrogen production. Alternatively, wastewaters with organic waste such as food processing waste have great potential as substrate sources.[13] Utilizing wastewaters from agricultural and food processing industries, which are generally high in carbohydrates, can provide the essential nutrients required for hydrogen production and reduce treatment and disposal costs currently needed for these particular waste streams. Removing and/or degrading waste materials and producing clean energy make anaerobic hydrogen production an attractive alternative to fossil fuels.[12]

Under normal anaerobic conditions hydrogen produced is consumed by methanogenic microbes.[14] To optimize hydrogen production, methanogens must be inhibited so that hydrogen producing bacteria can flourish. Excess hydrogen can be removed from the system as gas and used as an energy source. Aqueous system pH also has a profound effect on both hydrogen and other byproduct formation.[15] Given the importance of pH control to the system and the expense and safety concerns associated with chemical control, consideration should be given to use of substrates like food waste that maintain an optimal pH. Substrate composition and concentration are important issues that need to be understood and controlled.

At USU we have investigated hydrogen production using anaerobic fermentation. The overall objective was to investigate methods of producing hydrogen from agricultural and food wastes in a cost effective and efficient manner. One of the approaches to reducing cost was the use of the IBR anaerobic digester. By retaining and concentrating hydrogen producing bacteria in the reactor, this permits enhanced hydrogen production. It is designed to treat up to 12% influent solids including sand, stringy material and particles up to 38 mm in any dimension which may be found within food wastes.

Single phase and two phase digestion were investigated. The single phase involved producing hydrogen in batch mode CSTR bioreactors, while in the two phase experiments effluent from the hydrogen forming CSTR bioreactors was mixed together and fed into lab scale (56 L) IBR's where methane was made, separating the hydrogen forming and methane forming steps in the digestion process. The two phase experiments were comprised of a batch (CSTR) acidification step (HRT=7d) and continuously fed (IBR) methanogenesis step (HRT=4d).

2 METHODS FOR HYDROGEN EXPERIMENTS

2.1 Seed Preparation

The anaerobic hydrogen producing mixed microbial communities were enriched at the start of this study according to a method outlined in United States Patent 7540961. Seed sludge for the experiments was obtained from an anaerobic induced blanket reactor at the Wade Dairy Farm in Ogden, UT. Raw seed sludge was filtered through a screen (pore size: 2 mm) to remove large particles. The filtered raw sludge was preacidified at a pH of 3.0 and 37°C for 48 hours. After 48 hours, hydrochloric acid was mixed with the preacidified seed sludge for 10 min to lower the pH of the suspension to 2.0. Then the sample was stored at 4°C for 2 – 4 hours. The treatments inhibit the bioactivity of hydrogenotrophic non-sporeformers present in the manure.[16, 17] The enriched seed sludge was cultivated at 37°C, using a separate completely mixed batch reactor (working volume, 1.9 L) with an inoculation ratio of 30:70 for seed and glucose based substrate mixtures.

2.2 Medium Composition and Collection

Synthetic wastewater (SW), cheese whey, and manure were used in these experiments. SW consisted of organic and inorganic nutrients. It had a chemical oxygen demand (COD) of 25,000 mg/L, which was derived mainly from glucose. The solution was composed of approximately 21,300 mg/L glucose and the following nutrients (mg/L): 2,000 meat extract; 2,125 NH_4Cl; 420 K_2HPO_4; 180 $FeCl_2{\cdot}4H_2O$; 375 $CaCl_2{\cdot}2H_2O$; 312.5 $MgSO_4{\cdot}7H_2O$; 250 KCl. To prevent deficiency of microbial trace elements, a trace nutrients solution, (mg/L) 50 H_3BO_3, 50 $ZnCl_2$, 30 $CuCl_2$, 500 $MnSO_4{\cdot}H_2O$, 50 $(NH_4)_6Mo_7O_{24}{\cdot}4H_2O$, 50 $AlCl_3$, 50 $CoCl_2{\cdot}6H_2O$, 50 $NiCl_2$, and 1 mL HCl (36%) were added by 0.1% (v/v). The components were similar to those used for cultivating anaerobic bacteria.[18] 8,000 mg/L of $NaHCO_3$ was added to maintain initial buffering capacity. Tap water was used as diluting water (City of Logan, Utah).

Animal manure was collected from the Wade Dairy. Cheese whey was gathered at two cheese production plants (Gossner Cheese, Utah State Dairy, Logan, Utah). Plants produced primarily Swiss (Gossner) and Cheddar (USU) cheeses.

2.3 Reactors

Three Wheaton M-100 (Wheaton Instruments, Millville, NJ) CSTR lab scale bioreactors equipped with temperature controllers and magnetic agitation controls were used for the batch experiments. The anaerobic batch reactors were controlled at 37 ° C, and pH was maintained pH at about 5.5 by automatically feeding a 5 N mixed solution of NaOH if necessary using pH controllers (Cole-Parmer, Inc.). As a CSTR, HRT and SRT were equal (approximately 7 days or until the bioreactors quit producing hydrogen). Most hydrogen was produced within 48 hours. The head space was flushed with nitrogen gas prior to each trial. Water displacement was used to measure gas production.

Two phase hydrogen and methane production experiments were conducted (Figure 4) to compare total energy production from a single phase pilot scale IBR treating solely SW with the effluent from the hydrogen fermentation step which consisted of digested cheese whey, manure, and SW. All these were mixed together and stored at 0 °C before being added to pilot scale IBR's. The COD's of mixed effluent from the hydrogen production phase and the SW was 25 g/L for each. The feed tank and peristaltic pumps (Cole-Parmer, Inc., Vernon Hills, IL) were refrigerated at 2 °C when in operation to avoid degradation of the substrate during storage. The system was monitored for hydrogen and methane production.

Total COD in all experiments was measured by the closed reflux colorimetric method.[19] Total suspended solids (TSS) and volatile suspended solids (VSS) for biomass determination were analyzed and calculated from influent and effluent samples.

The hydrogen, methane, oxygen, and nitrogen contents in the biogas were analyzed by gas chromatography (HP 6890 series, Hewlett-Packard, Wilmington, DE) using an RT-Msieve 5A Plot capillary column (Restek) with dimensions of 30.0 m * 320 μm * 30.0 μm. The column temperature was 35 °C, while the inlet port and thermal conductivity detector temperatures were 43 °C and 200 °C, respectively. Argon was used as the carrier gas at a flow rate of 3.3 mL / min. Gas standards were obtained from Scott Specialty Gases (Plumsteadville, PA). Samples of methane (99.0%), nitrogen (80.0%), and hydrogen (10.0%) were used in calibrating the gas chromatograph.

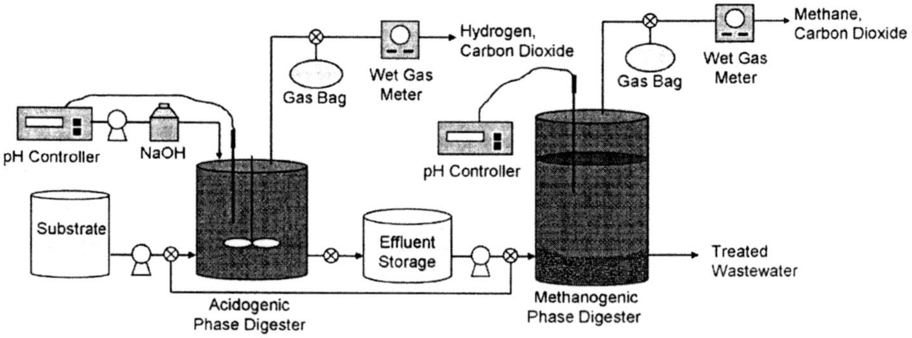

Figure 4 *Process flow diagram for two stage anaerobic production of hydrogen and methane using cheese whey, synthetic waste, and manure.*

2.4 Experimental design

The enriched seed sludge was added to the anaerobic batch reactors at an inoculation ratio of 1:9 for seed and substrate mixtures for trial one. The subsequent trials used effluent from the previous trial to seed the next trial at a ratio of 1:9 in accordance with previous related experiments.[17]

The first trials used synthetic waste (SW) as substrate for the hydrogen fermentations. The main purpose of these trials was to determine optimal pH within a range of 5.0 -6.0 and quality and quantity of biogas production.[15, 20]

The manure and cheese whey trials were mixed with SW at concentrations of 0, 15, 30, and 45% cheese whey with three trials per concentration being tested. The manure trials contained an extra trial of 100% manure. The cheese whey and manure trials were similar in that cheese whey was mixed with manure at concentrations of 0, 15, 30, and 45% cheese whey.

It was determined in preliminary experiments that effluent from the hydrogen production process would produce methane when added to an IBR. Trials were run to quantify how much methane would be produced and how COD would be removed in a single-phase anaerobic digestion (SPAD) process compared to the methanogenic phase of a two-phase anaerobic digestion (TPAD-MP) that produced hydrogen in the first phase and methane in the second. IBR's were used for both the SPAD and TPAD-MP methane fermentations. The IBR's used for these trials ran on a continuous 4-day hydraulic retention time. Synthetic wastewater was utilized as substrate for the SPAD IBR anaerobic reactions. The TPAD-MP reactions were performed with a concentration of 50% effluent from the first phase hydrogen production and 50% synthetic wastewater. Effluent from all the hydrogen single phase experiments was mixed together and stored at 0 °C before being used in the TPAD-MP trials.

3 RESULTS

A lag phase was noted at the beginning of the single phase hydrogen trials ranging between 2-4 hours after adding the various substrates. The majority of the biogas production was produced within 24 hours after inoculation with the stress enriched bacterial culture.

3.1 Synthetic Wastewater Trials

Preliminary tests without pH control showed the system becoming increasingly acidic within 4-6 hours. The pH dropped to values of 3 - 4. The system produced significant amounts of biogas (hydrogen & carbon dioxide) until dropping below pH 5. At this point biogas production decreased rapidly (data not shown).

Three trials were successfully performed using the synthetic wastewater and controlling the pH at 5.0, 5.5, and 6.0. Synthetic wastewater trials produced 1.36 liters of hydrogen per liter of substrate at a pH of 5.5, (Figure 5). This was the best pH in terms of volume of hydrogen produced and inhibition of methanogenic bacteria. The average hydrogen percentage within the biogas at pH 5.5 was 40%. The chemical oxygen demand tests showed a COD removal of 19 % ± 3.7 % at pH 5.5.

3.2 Cheese Whey and SW Trials

Trial one gave very promising data showing high hydrogen volumes (Figure 5). The 45% cheese whey concentration produced 2.0 liters of hydrogen per liter substrate (all biogas volumes in the results section are reported at STP). The hydrogen concentration within the biogas ranged between 28 - 39%. The second trial using 15 and 30% cheese whey concentrations with synthetic wastewater produced significantly less biogas with no biogas formation in the 45% concentration. The third trial with cheese whey concentrations of 15, 30, and 45% did not produce biogas. Biogas production dropped off significantly and COD removal stayed consistently high. The 45% cheese whey concentration had an average removal of 13.6 grams per liter or 33.2% total COD removal. Similar results were reported by research in which an increase of lactic acid producing bacteria within the digestion was blamed for lower biogas yields.[21] Results from the current research and conclusions from previous research suggest that the use of cheese whey and poor mixing in the reactor favored lactic acid bacteria growth, which out competed the hydrogen producing bacteria causing a decrease in hydrogen production.

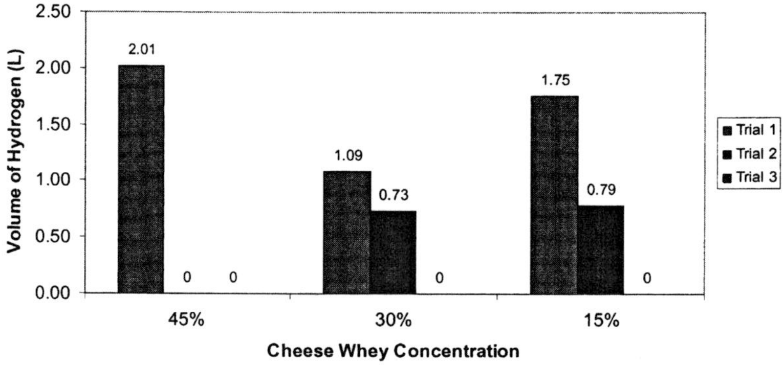

Figure 5 *Results of hydrogen production (STP) for each of the cheese whey mixed with synthetic wastewater trials. Trial three showed no hydrogen formation and trial two showed reduced hydrogen production. Results are reported as volume of hydrogen per liter substrate.*

The average pH of the cheese whey used in the digestions was 6.4. Initially, cheese whey was stored at -20 °C until used for the trials. After further investigation, it was decided to submit the cheese whey to a proprietary process that made it more suitable for hydrogen manufacture. The pretreated cheese whey was then utilized in the cheese whey and manure trials.

3.3 Cheese Whey and Manure Trials

Three trials were completed using pretreated cheese whey and manure. It was observed that the higher the pretreated cheese whey concentration, the more hydrogen produced. The 45% pretreated cheese whey mixed with manure produced 1.6 liters of hydrogen per liter substrate. The average hydrogen content of the biogas at this concentration was 36 %. COD removal was satisfactory (data not shown here).[22] The pretreated cheese whey had several advantages when utilized as a substrate. The pretreatment process caused a more stable digestion process by consistently producing a significant and constant amount of hydrogen at every concentration tested. Another major advantage was that the low pH of the cheese whey caused the pH of the total substrate to drop to an inhibitory range for methanogenic bacteria, while still providing nutrients for the hydrogen producing bacteria. The low pH of the cheese whey eliminated the need for pH control to promote hydrogen production. The 45% pretreated cheese whey mixed with manure produced 16.0 kJ/L of energy.[22]

3.4 Methane Production Trials

An average of 81.07 ± 12.76 mmol per day of methane per liter substrate was produced in the methanogenic phase of the two phase reactor (Table 1). This was compared against the single-phase reactor production rate of 72.72 ± 11.93 mmol per day of methane per liter substrate. Another advantage observed during the methanogenic phase trials was the amount of chemically oxidizable organic matter removed. Removal rates of 59.44% were observed in the methanogenic phase reactor compared to the single-phase reactor removal of 44.05%.

Table 1 *Summary of results of the hydrogen and methane tests.*

	Trial	COD Removal (%)	Biogas[a] (L/L substrate)	Hydrogen[a] (mmol/L)	Methane[a] (mmol/L)	Energy Potential (kJ/L)
Hydrogen Tests	0% CW 100% Man	10.24%	0.26 ± 0.16	0.41	0	0.1
	15% CW 85% Man	4.36%	2.18 ± 0.12	25.77	0	6.3
	30% CW 70% Man	7.83%	3.95 ± 0.61	52.49	0	12.9
	45% CW 55% Man	12.46%	4.37 ± 0.66	63.16	0	15.55
Methane Tests[b]	SPADc	44.05%	2.95 ± 0.42	0.81 ± 0.60	72.72 ± 11.93	58.51
	T PAD-MP[d]	59.44%	2.84 ± 0.47	0.23 ± 0.32	81.07 ± 12.76	64.99

[a]At standard temperature and pressure
[b]Based on a per day average
[c]Single phase anaerobic digester
[d]Two phase anaerobic digester-methanogenic phase

4 CONCLUSIONS

Fermentation of cheese whey, glucose based medium, and animal manure to produce hydrogen gas was successful using seed treated to inhibit bioactivity of hydrogenotrophic microorganisms. Hydrogen production using pretreated cheese whey and manure as substrate was greatest for the pretreated cheese whey and manure mixture. All the effluents from hydrogen production subsequently produced methane in an IBR anaerobic digester.

The total energy produced as biogas and COD removal was better with a two phase system (H_2 in first phase and CH_4 in second) compared to single phase production of CH_4 for the substrates tested.

References

1 G. Lettinga, A.F.M. v. Velsen, S.W. Hobma, W. d. Zeeuw and A. Klapwijk, *Biotechnol. Bioeng.*, 1980, **22**, 699.
2 M.H. Gerardi, The microbiology of anaerobic digesters, Wiley-Interscience, Hoboken, N.J., 2003.
3 G. Tchobanoglous, F.L. Burton, H.D. Stensel and Metcalf & Eddy., Wastewater engineering : treatment and reuse, 4th edn., McGraw-Hill, Boston, 2003.
4 O. Levenspiel, Chemical reaction engineering, 3rd edn., Wiley, New York, 1999.
5 P.V. Danckwerts, *Chem. Eng. Sci.*, 1995, **50**, 3857.
6 J.S. Dustin and C.L. Hansen, *Trans. ASABE*, 2009, 2009.
7 M. M. Zurawski D., Stegmann R., in Sardinai 2005, Tenth International Waste Management and Landfill Symposium, CISA, Environmental Sanitary Engineering Centre, Italy, S. Margherita di Pula, Cagliari, Italy, Editon edn., 2005.
8 G. Antonopoulou, H. N. Gavala, I. V. Skiadas, K. Angelopoulos and G. Lyberatos, *Bioresour. Technol.*, 2008, **99**, 110.
9 J.H. Jo, C.O. Jeon, D.S. Lee and J.M. Park, *J. Biotechnol.*, 2007, **131**, 300.
10 K. Nath and D. Das, *Appl. Microbiol. Biotechnol.*, 2004, **65**, 520.
11 J.J. Romm, The hype about hydrogen : fact and fiction in the race to save the climate, Island Press, Washington, DC, 2004.
12 I.K. Kapdan and F. Kargi, *Enzyme Microb. Technol.*, 2006, **38**, 569.
13 J. Benemann, *Nat. Biotechnol.*, 1996, **14**, 1101.
14 W. Parawira, M. Murto, R. Zvauya and B. Mattiasson, *Renewable Energy*, 2004, **29**, 1811.
15 S.K. Khanal, W.-H. Chen, L. Li and S. Sung, *Int. J. Hydrogen Energy*, 2004, **29**, 1123.
16 C.C. Chen, C.Y. Lin and M.C. Lin, *Appl. Microbiol. Biotechnol.*, 2002, **58**, 224.
17 D.-Y. Cheong and C.L. Hansen, *Bioresour. Technol.*, 2007, **98**, 2229.
18 A.J.B. Zehnder, B.A. Huser, T.D. Brock and K. Wuhrmann, *Arch. Microbiol.*, 1980, **124**, 1.
19 A.E. APHA, M. Franson, American Water Works Association, Standard methods for the examination of water & wastewater, 21 edn., American Public Health Association, 2005, 2005.
20 S.V. Ginkel, S. Sung and J.-J. Lay, *Environ. Sci. Technol.*, 2001, **35**, 4726.
21 M. Cooney, N. Maynard, C. Cannizzaro and J. Benemann, *Bioresour. Technol.*, 2007, 98, 2641.
22 R.S. Thompson, M.S. Thesis, Utah State University, Logan, UT 84322 2008, 74 p.

THE EFFECT OF ALCOHOLS ON CELLULASE ACTIVITY

A. Elliston,[1,2] C.B. Faulds [1] C.J. Barry[2] and K.W. Waldron[1]

[1] Institute of Food Research; Sustainability of the Food Chain Exploitation Platform, Norwich NR4 7UA, UK
[2] Achor Innovation Limited, Colney Hall, Norwich NR4 7TY, UK

1 INTRODUCTION

The increased production of biofuel in recent years can be explained by a number of environmental, political and economic reasons. Firstly, the Kyoto protocol[1] requires that the 183 countries that make up the United Nations Framework Convention on Climate Change (UNFCCC) are legally committed to reducing greenhouse gas emissions. Secondly, more localised legislation such as the Renewable Transport Fuel Obligation in the UK requires that 5% bioethanol produced from a renewable source has to be added to all fuel sold at the pump. Thirdly, the escalating price of crude oil has provided a direct economic basis to stimulate the rapid rise in biofuel production over the last ten years.[2]

Biofuels have predominantly been produced from food crops such as wheat and sugar cane. However whilst these can be considered by some to be a useful resource, their exploitation in this way has several drawbacks. Using food crops to create bio-fuel means that a competitive market is created between food and fuel manufacturers thereby creating inflated prices for these crops. Also, the inherent seasonality of many crops means that the biofuel industries are not always able to operate all the year round. Furthermore, only relatively small percentages of the crop plants can be used in the so-called "First generation" biofuel production. "Second generation" biofuel processes are now shifting towards using lignocellulosic biomass as their base resource.

The second generation biofuel processes convert cellulose-containing plant biomass into ethanol in four main steps. 1) Pre-treatment: to improve accessibility of hydrolytic enzymes to the cellulosic substrate. 2) Hydrolysis: the conversion of polysaccharides to soluble sugars. 3) Fermentation: glucose and other usable sugars are then fermented to produce alcohols; yeast is generally used to create ethanol but other micro-organisms such as clostridia can be used to produce higher-chain alcohols, such as butanol. 4) Distillation: finally the alcohol is distilled to concentrations that are commercially usable.

Agro-industrial by-products are beginning to be seriously considered as potential resources. A large proportion of food, paper and green waste have the potential to be used to some degree in the production of second generation biofuels. With millions of tonnes of this resource still being sent to landfill each year in the UK alone, which cost industry and the government financially as well as resulting environmental issues, this could easily become one of the biggest feed stocks for biofuel production in coming years.

In the interest of efficiency, simultaneous saccharification and fermentation (SSF) is becoming more commonly considered for bioalcohol production. SSF combines steps 2) and 3) above so that they are conducted simultaneously in the same vessel. This enables the fermenting microorganism to use the sugar directly as it is produced by the enzyme hydrolysis, overcoming end point inhibition of the enzyme that would otherwise result in a lower efficiency. Moreover production of alcohol helps prevent microbial contamination of the process and combining the two vessels decreases initial plant cost.

The development of commercial enzymes to hydrolyse lignocellulosic biomass has increased rapidly in recent years. Proprietary cellulase enzyme mixes are now being specifically formulated with improved yield and reaction times in mind for these processes. This paper discusses the quantitative inhibition of enzymatic hydrolysis action by alcohols at concentrations that may be produced within the SSF process. With the current interest in biofuels, there is much debate regarding the most effective alcohol to use in biofuel formulation. All present a range of benefits and disadvantages. Thus, the study considers the variation in inhibitory effects due to the carbon chain length of the alcohols.

2 METHOD AND RESULTS

2.1 Method

Four proprietary cellulase mixtures, Acid Cellulase (NBS Biologicals, Huntingdon, UK); Accellerase® 1000 (Genencor, Rochester, N.Y., USA); Depol 740L and 13L-CO13L (Biocatalysts Ltd, Cardiff, UK), chosen on the basis of high cellulase activity, were assayed in the presence of methanol, ethanol, propan-2-ol and n-butanol for the hydrolysis of Pure Sigmacell® Cellulose, type 20, and Whatman No 1 filter paper circles (6mm diameter). These alcohols were chosen as they are the most widely accepted petro-diesel substitutes and therefore the most likely to be produced in an SSF process. Hydrolytic assays (250 μl) in 100 mM sodium acetate (pH 5.0) were performed using a thermocirculating incubator (50 °C) and samples removed every 30 minutes. The 2,4-dinitrosalicylic acid (DNS) colourimetric assay[3] for reducing sugars was used to determine total reducing sugars within the saccharification liquor at the end of the time period. The amount of sugar released was determined against a standard curve of glucose.

2.2 Results

In Figures 1-4 below, inhibition is measured as the percentage change compared to the test sample in the absence of alcohol. Although all four enzymes were tested in this experiment, only the results for Depol 740L are represented for simplicity, while the other enzymes behaved in a similar manner.

2.2.1 General Inhibition. Over an alcohol concentration range of 0-10% (v/v), a common trend was observed in which increasing the concentration of alcohol resulted in an increase in the degree of inhibition of the measured activity. The results demonstrate that concentrations above approximately 3% (v/v) alcohol have a significant inhibitory effect on the action of the saccharification and would therefore have severely attenuated the SSF process as concentrations increase.

2.2.2 Activation at low alcohol concentrations. Conversely it can also be seen (predominantly in Figure 2 and to a lesser extent Figure 4) that there is an activation effect on hydrolytic activity at alcohol concentrations of approximately 2% (v/v). This is less obvious in Figures 1 and 3 but there is a visible inflection on the graphs again at 2% (v/v) concentration giving less than expected inhibition rather.

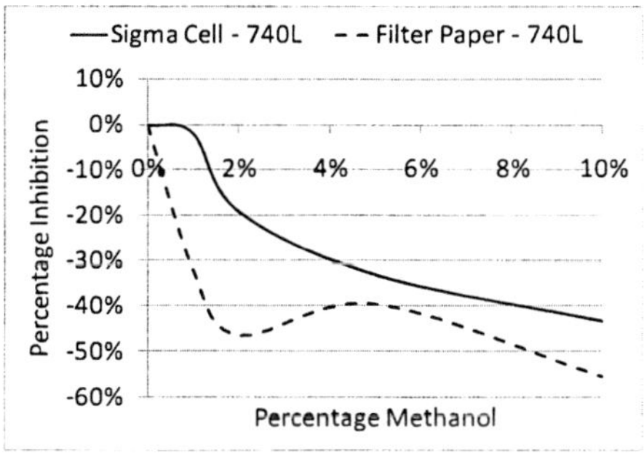

Figure 1 *Methanol Inhibition*

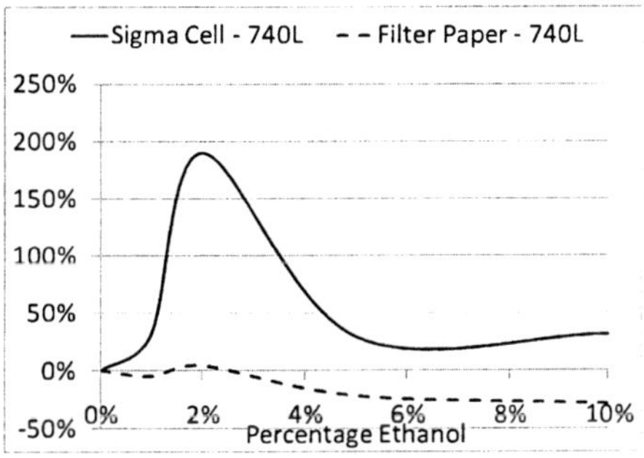

Figure 2 *Ethanol Inhibition*

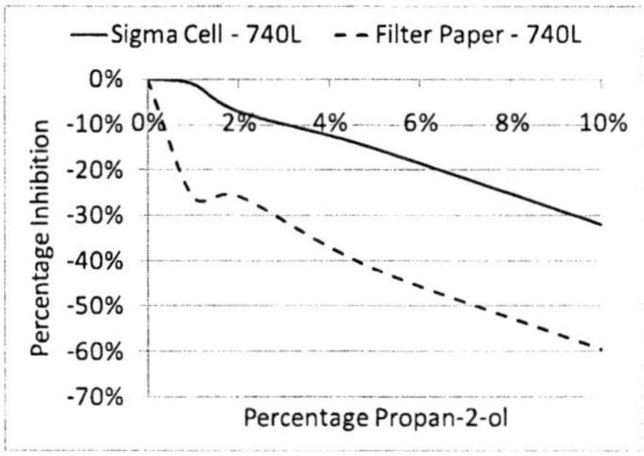

Figure 3 *Propan-2-ol Inhibition*

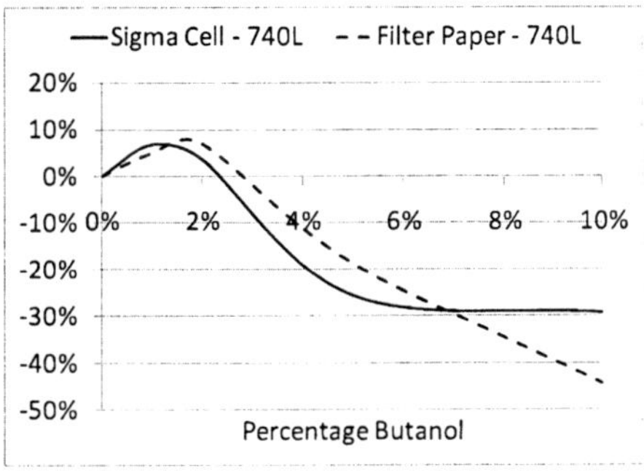

Figure 4 *n-Butanol Inhibition*

In order to shed more light on the impact of alcohols at low concentrations, a detailed study was carried out to evaluate more precisely the concentration at which alcohols might have the greatest stimulatory effect. As ethanol was found to give the greatest effect on Sigmacell® this was chosen as the substrate. A larger number of sample points were taken between 1 and 5% ethanol and the results can be seen in Figure 5. It can be seen that ethanol has an optimal activation effect at 2.4% (v/v), and that above 4% it starts to have a negative effect.

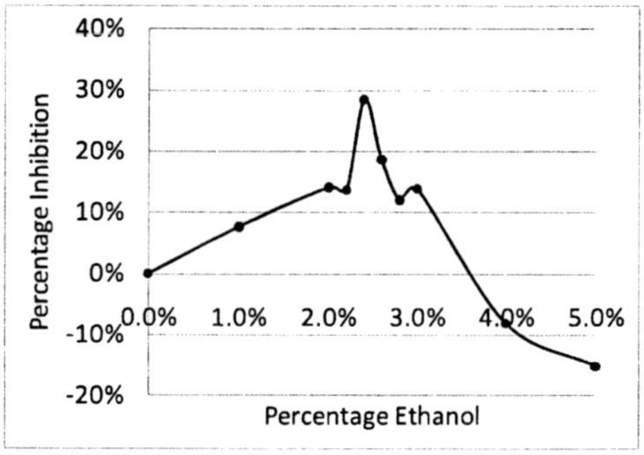

Figure 5 *Activation optimisation of Depol 740L with Ethanol*

2.2.3 Solubility and Detergency. It is possible that the activation effect of alcohol at low concentrations is due to an increase in substrate solubility, and/or an increase in the accessibility of the enzyme to the substrate, thereby increasing the efficiency of saccharification. To test this hypothesis a proprietary detergent was used in place of alcohol. When Tween 20 was used in place of alcohol (Figure 6), the detergent had strikingly similar results to that seen with ethanol (Figure 2). Therefore, this would seem to suggest that a hydrophobicity effect may underlie the stimulation of enzyme saccharification.

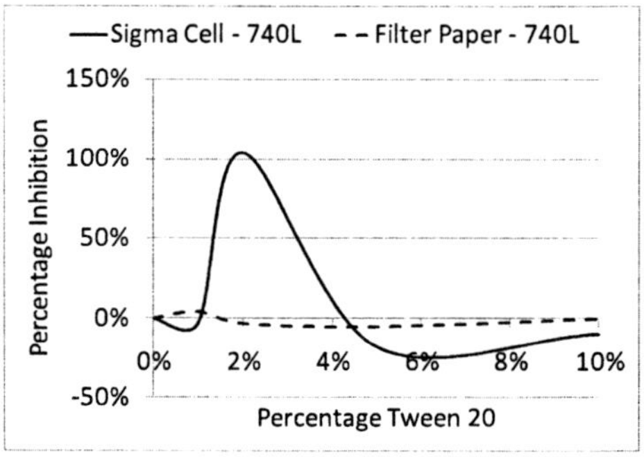

Figure 6 *Tween 20 Inhibition of cellulose activity*

3 CONCLUSION

Increasing quantities of alcohol in the saccharification process leads to a general inhibition of the cellulase enzymes. There seems to be little effect of alcohol chain length on the extent of inhibition produced. However there also appears to be a very marked activation effect at low concentrations (2-3% v/v) of alcohol, most notably by ethanol. This activation effect may have considerable positive effects on SSF processes used in bio-fuel manufacture. However the concentration of alcohol would have to be closely monitored and controlled in order to avoid inhibition of hydrolysis at higher concentrations.

This stimulatory effect may be caused by the solvation characteristics of the alcohols used in the experiments. Neither Sigmacell® or the filter paper used are fully soluble and the change in water content through the addition of alcohol or detergent would appear to increase the accessibility of the cellulases to the substrates (more so in the Sigmacell®).

Acknowledgements

The Authors gratefully acknowledge the support of the UK Biotechnology and Biological Science Research Council and Achor Innovation Limited in funding this CASE studentship.

References

1 United Nations, Editon edn., 1998.
2 Earth Policy Institute, *World Ethanol Production, 1975 - 2005*, http://www.earth-policy.org/Updates/2006/Update55_data.htm.
3 J.B. Sumner and V.A. Graham, *J. Biol. Chem.*, 1921, **47**, 5.

WATER RECYCLING AND RECOVERY IN FOOD AND DRINK PROCESSING

Jiří Klemeš[1], H.L. Lam[1] and D.C.Y. Foo[2]

[1]Centre for Process Integration and Intensification – CPI[2], Research Institute of Chemical and Process Engineering, Faculty of Information Technology, University of Pannonia, Egyetem utca 10, Veszprém, H - 8200, Hungary
[2]Department of Chemical and Environmental Engineering, University of Nottingham Malaysia, Broga Road, 43500 Semenyih, Selangor, Malaysia

1 INTRODUCTION

Water is widely used in the food and drink industries as an important raw material. It is also frequently used in the heating and cooling utility systems, e.g. steam production, cooling water, as well as being a mass separating agent for various mass transfer operations (e.g. washing or extraction). Strict requirement for product quality and associated safety issues in manufacturing contribute to large amounts of high-quality water being consumed by the food industry. Furthermore, stringent environmental regulations as well as growing human population with improved quality of lifes have led to growing demands for good quality water. These changes have increased the need for improved water management and wastewater minimisation. The adoption of water minimisation techniques can effectively reduce both the fresh water demand and subsequently the effluent generation in the food industry. This results in the reduced cost for fresh water acquisition and effluent treatment.

Hoekstra [1] defined the *water footprint* (WFP) as an indicator of direct and indirect water use, which is measured in terms of water volumes either consumed, evaporated and/or polluted. The WFP includes consumptive use of virtual *green*, *blue* and *grey water*. The virtual green water content of a product is the volume of rainwater that evaporated during the production process. For the food industry, it is mainly consumed by agricultural products, where it refers to the total rainwater evaporation from the field during the growing period of a crop. The virtual blue water content of a product is the volume of surface water or groundwater that evaporates as a result of production. An example includes the evaporation of irrigation water from a field, irrigation canals and storage reservoirs. The virtual grey water content of a product is the volume of water required to dilute pollutants to meet the water quality standards for reuse or discharge to the environment.

A water footprint can be calculated for any product or activity as long as the group of consumers and producers are well-defined. The water footprint is a geographically and temporally explicit indicator, not only showing volumes of water consumption and pollution, but also referring specifically to the type of water use, as well as where and when the water was used.

The idea of water LCA has gained more interest after the introduction of the water footprint concept.[1,2] Considering water use along food supply chains, the real water content of the product is generally negligible if compared to the virtual water content, which is the total fresh water being used at the various steps of product supply/production chains. Apart from being used as an ingredient in food preparation, most of the water use in food industry is categorised as virtual water.[3] The most commonly found water-using operations include the following:

i. Heating – boiler, heat exchangers, etc
ii. Process water – cooling towers
iii. Potable uses – offices, canteens, etc
iv. Washing – equipment, bottles, floor, vehicles, etc
v. Rinsing – equipment, bottles, food materials, final products
vi. Fire fighting
vii. Transport medium

The food industry is characterised as a high water consumer. For example, the annual total water consumption for the food industry was estimated as 347.2 Mm^3 in Canada [4] and 455 Mm^3 in Germany.[5] Some recent studies [6-7] have reported the virtual water for the production of some common food products. The findings are summarised in Table 1.

Table 1 *Average virtual water content of some selected food products* [7].

Product	Virtual water (L)
1 glass of beer (250 mL)	75
1 glass of milk (200 mL)	200
1 cup of coffee (125 mL)	140
1 cup of tea (250 mL)	35
1 slice of bread (30 g)	40
1 slice of bread (30 g) with cheese (10 g)	90
1 potato (100 g)	25
1 apple (100 g)	70
1 glass of wine (125 mL)	120
1 glass of apple juice (200 mL)	190
1 glass of orange juice (200 mL)	170
1 bag of potato crisps (200 g)	185
1 egg (40 g)	135
1 hamburger (150 g)	2,400
1 tomato (70 g)	13
1 orange (100 g)	50

In the food industry, it is advisable to approach water recovery from the simplest measures based on efficient management (e.g. good housekeeping) to more advanced methodologies (e.g. process integration techniques). Food processing is not always operated in continuous mode, as it is intermittent and highly dependent on the availability of the feed stock. Typical production campaigns occur in sugar, fruit juice or cereal processing plants. On the other hand, breweries are in operation on a continuous basis, but are batch processed. These features affect any technologies to be adopted, including those involving water and waste minimization and processing.

A more comprehensive summary of water and wastewater minimisation application for the food industry can be found in the books of Klemeš, Smith and Kim [8] and Smith [9]:

i. Process changes – this enables the fundamental changes on unit operations that consume fresh water. This includes the increase of the number of stages in an extraction process to reduce its inherent water use, changing from wet cooling towers to air coolers, improved energy efficiency to reduce steam demand, increased condensate return from steam systems, as well as good housekeeping. For the latter, analysing and measuring water use and wastage, reducing water wastage, cleaning operations, maintenance of equipment and daily work practices are some good practises.
ii. Reuse/recycle/regeneration – these options enable the recovery of wastewater in water-consuming operations. However, the presence of pollutants in the wastewater streams needs to be taken into consideration in order not to affect the water-using processes. This aspect will be discussed in greater detail in the following section.

Important methodologies for water minimisation are listed by Klemeš and Perry [10]:

- The use of *water pinch analysis* techniques.
- The use of *mathematical optimisation* techniques.
- Minimisation of water consumption by efficient management and control of process operations.
- The integration of optimisation and production planning techniques in conjunction with real-time plant measurements and control for product quality and minimisation of losses.

This paper demonstrates two different case studies, utilising the first two methods on the above list. Note that both the water pinch analysis and mathematical optimisation techniques belong to the systematic framework of *process integration*. One of its definitions is given as "a holistic approach to process design, retrofitting and operation which emphasises the unity of the process".[11] Water pinch analysis and mathematical optimisation techniques supplement each other well. While the former locates various network targets prior to detailed design; the latter addresses more complex systems, e.g. multiple impurities, cost considerations, topological and process constraints.

2 WATER MINIMISATION WITH WATER PINCH ANALYSIS TECHNIUQUE

2.1 Introduction to Water Pinch Analysis

Pinch analysis was initiated for heat exchanger network synthesis, and subsequently has been extended into other energy integration works.[9,12,13] Based on the analogy of heat and mass transfer, it was then extended for the synthesis of mass exchange network and a series of other mass integration applications.[12] Water pinch analysis then emerged as a special case of mass integration, following the seminal work of Wang and Smith.[14] However, their targeting technique was limited to the *fixed load problem* where water-using processes are modelled as mass transfer operations. Later work on water pinch analysis has been mainly dedicated to the *fixed flowrate problem* where flowrate requirement is the taken as the primary concern for water-using processes.[15,16,17,18,19]

In the context of water pinch analysis, *reuse* means that the effluent from one unit is used in another unit and does not re-enter the unit where it has been previously used. On the other hand, *recycle* allows the effluent to re-enter the unit where it has been previously used. Besides, one may also utilise a regeneration unit (e.g. filter, stripper, etc.) to partially purify the water stream prior to reuse/recycle.[14]

A typical pinch analysis study consists of two stages, i.e. setting minimum fresh water and wastewater flowrates (often termed as *targeting*), followed by network design to achieve the targeted flowrates. It is worth mentioning that the focus on water pinch analysis is the targeting step. With targeting, we are able to set a baseline target to determine how well a reuse/recycle system can actually perform based on first principle. Knowing the targets ahead of design will eliminate the query of "will there be a better design?" during any design exercise. Once the minimum material targets are established, a water network can be designed using any network design tools.

To set the flowrate targets for water reuse/recycle, various graphical and tabulated targeting techniques may be employed. This includes the limiting composite curves,[14] water surplus diagram,[16] material recovery pinch diagram,[17,18] cascade analysis technique [19] and source composite curve.[20] After the flowrate targets are identified, numerous techniques can be used to design a water network that achieves the flowrate targets. Note that above-mentioned works were developed for continuous processes. Several water pinch analysis works dedicated for batch processes have also been reported. This includes the works of Wang and Smith,[21] Foo *et al.*,[22] Liu *et al.*[23] and Majozi *et al.*[24]

2.2 Flowrate Targeting with MRPD

In the part the targeting technique of material recovery pinch diagram, i.e. MRPD [17, 18] is illustrated. To construct a MRPD, material flowrates and loads of the individual process sink and source are needed. The steps required to construct the MRPD are given as follows: [17,18]

i. Arrange the water sink (SK_j)/source (SR_i) individually in ascending order of concentration level (C).
ii. Calculate the load for each sink and source, given by the product of its flowrate (F) and concentration level.
iii. Plot the sink/source on a load-versus-flowrate diagram with an ascending order of their concentration levels (given by the slope of the plot). Connect the tail of the individual sink/source segments to the arrowhead of another sink/source streams to form the sink/source composites.
iv. For pure fresh resources (with zero impurity concentration), the sink and source composite curves are separated horizontally until they just touch, with the source composite lying below and to the right of the sink composite, as shown Figure 1(a).
v. For impure fresh resources, the source composite curve is shifted along an impure fresh locus until it lies below and to the right of the sink composite, as shown Figure 1(b).

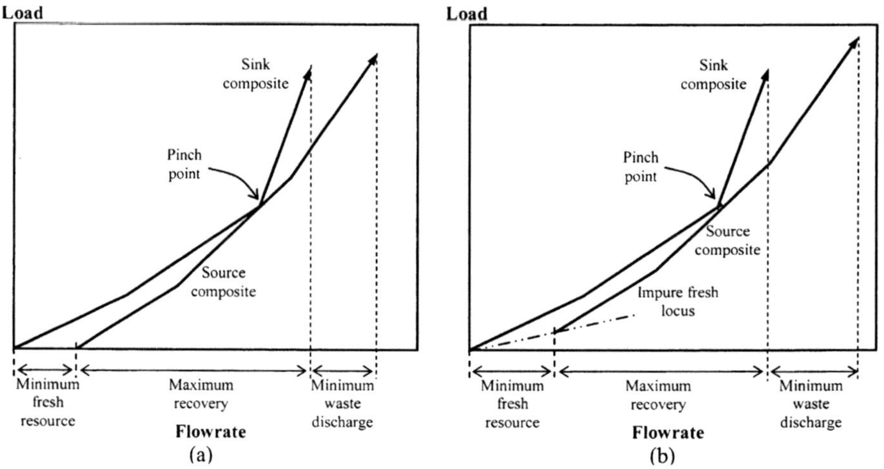

Figure 1 *MRPD for: (a) Pure fresh resource (no impurity); (b) Impure fresh resource*

The overlap area of the sink and course composites represents the maximum recovery among all sink and source streams within the network. The point where the two composites touch termed as the material recovery pinch, which is the bottleneck for maximum recovery. The segment where the sink composite extends to the left of the source composite represents the minimum feed needed for fresh resources (to be purchased); while region where the source composite extends to the right of the sink composite represents the minimum waste discharge from the network (for final treatment before discharging to the environment).

The minimum fresh resource needed and the minimum waste generated by the network both constitute the resource targets of the network, and are determined ahead of detailed design of the recovery network. In the following section, the MRPD is used to locate the flowrate targets for a fruit juice production case study.

2.3 Demonstration Case study: Fruit juice production

Table 2 shows the limiting water data for a fruit juice production case study.[25,26] The original case study was operated in batch mode. Each water sink and source has its start (t^S) and end time (t^T). Prior to water recovery, freshwater and wastewater flows for this case study are calculated as 96 m^3 and 84 m^3 (given by the sum of the individual water flow).

Table 2 *Limiting water data for Fruit juice production case study.*

SK_j	$F_{SKj}(m^3)$	C_j (ppm)	t_j^S (h)	t_j^T (h)	SR_i	$F_{SKi}(m^3)$	C_i (ppm)	t_i^S (h)	t_i^T (h)
SK1	20	0	0.5	2.5	SR1	20	5	2.5	4.5
SK2	20	6	5.0	7.0	SR2	20	14	7.0	9.0
SK3	20	15	9.5	11.5	SR3	20	20	11.5	13.5
SK4	16	5	17.0	19.0	SR4	8	25	17.0	19.0
SK5	20	7	6.0	8.0	SR5	16	10	10.5	14.5

To simplify the analysis, the following assumptions are made:

i. The batch process is operated repeatedly on a yearly basis. As a consequence it behaves as if it is operated in continuous mode. Earlier works [22] proved that a repeated batch operation will always achieve the flow targets as in the continuous mode.
ii. An unlimited water storage tank is always available for storing water for later use.
iii. Water recovery is always carried out between two consecutive batches. In other words, water sources in an earlier batch will be sent to water storage tanks, before it is reused/recycled to the water sinks in next batch operation. Similar assumption was made in an earlier work.[27]

Given the above assumptions, any established targeting technique for continuous process can be utilised to identity rigorous water targets for the case study. For this case, the MRPD is used. The MRPD for the case study is shown in Figure 2(a); while the network that achieves the water targets is shown in Figure 2(b).

As shown, the minimum fresh water flow (F_{FW}) is determined as 35 m^3; while the wastewater flow (F_{WW}) is 23 m^3. Compared to the total water flow prior to water recovery, this represents a significant flow reduction of 63.5 and 72.6% for fresh water and wastewater respectively.

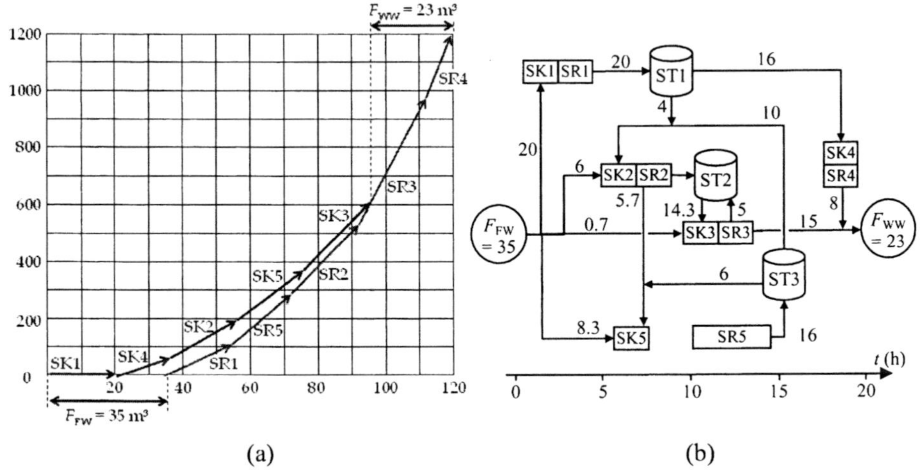

(a) (b)

Figure 2 (a) *Material recovery pinch diagram;* (b) *Network design for fruit juice production case study*

3 WATER MINIMISATION WITH MATHEMATICAL OPTIMISATION APPROACH

3.1 Introduction to Mathematical Optimisation Approach

Apart from water pinch analysis, mathematical optimisation techniques have also been utilised for water minimisation. Various mathematical optimisation approaches have been developed to complement water pinch analysis in dealing with more complex problems, e.g. multi-contaminant systems [28,29,30] complex operational constraints which include limiting the number of pipeline connections [31] and forbidden/compulsory matches between water-using processes.[32,33,34] Some more recent research results and case studies have been published recently by various authors from South Africa,[35,36] Asia [37,38,39] and Europe.[40]

3.2 Demonstration Case Study: Brewery Plant

A case study on a brewery plant [40] is briefly discussed to illustrate the use of a mathematical optimisation technique for water minimisation. In the brewery case study, the volume ratio of water consumption to beer sold was reported as 6.04 L/L or 653,300 m^3/y. Based on the ratio set by BREF [41] the freshwater consumption exceeded the upper limit by 144,900 m^3. For this reason an initiative for the process improvement emerged within the company for retrofit of the existing water network to minimise fresh water usage of the plant.

However, the process consists of a mixture of batch and semi-continuous water-using processes. Water-using operations in the packaging area are operated mainly in batch mode with the exception of rinsers for non-returnable glass bottles and cans. Wastewater stream from semi-continuous processes can be reused in batch processes with lower purity requirement. Motivated with this opportunity, the basic formulation is firstly proposed to enable efficient integration of semi-continuous and batch water-using processes.

The continuous wastewater streams are treated as limited freshwater sources and the unused wastewater is discharged. In the next step the model is further extended by the option of installing intermediate storage tanks for the collection of those unused wastewater streams from semi-continuous processes to be recovered over the subsequent time intervals. The reason for this model extension is the difference in operating schedule of the filling lines. The superstructure representation for water reuse and regeneration reuse for the case study is shown in Fig. 3.

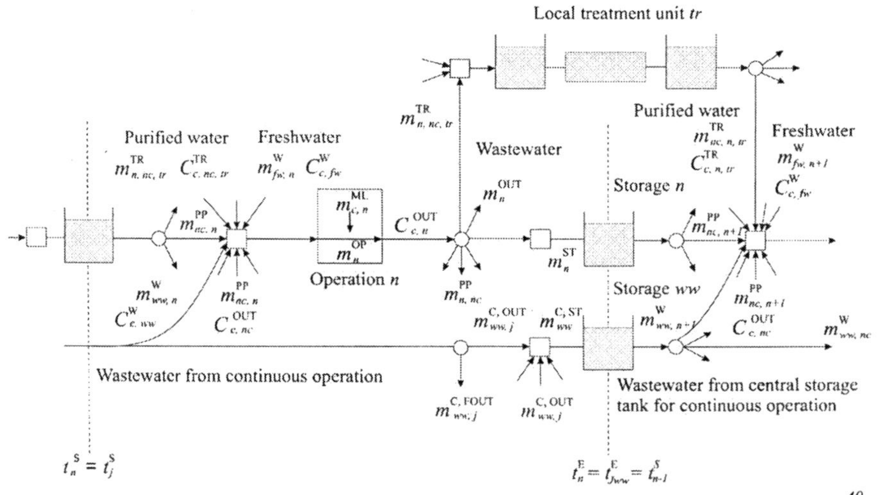

Figure 3 *Superstructure for water re-use and regeneration re-use in a brewery plant* [40]

Opportunities for regeneration reuse were analysed in the brew house and the cellar because of high contaminant concentration at these sites (see Fig. 4). In this case the basic model is extended with the option of installing a local (on-site) wastewater treatment unit operating in batch or in continuous mode. The extended model then enables water regeneration reuse/recycle. Scheduling of batch wastewater treatment units is performed simultaneously in order to adjust the treatment schedule with the fixed schedule of batch process. The option of storage tank installation before and after treatment is included in the model in order to store wastewater and/or purified water if the time schedule of the treatment unit requires storage.

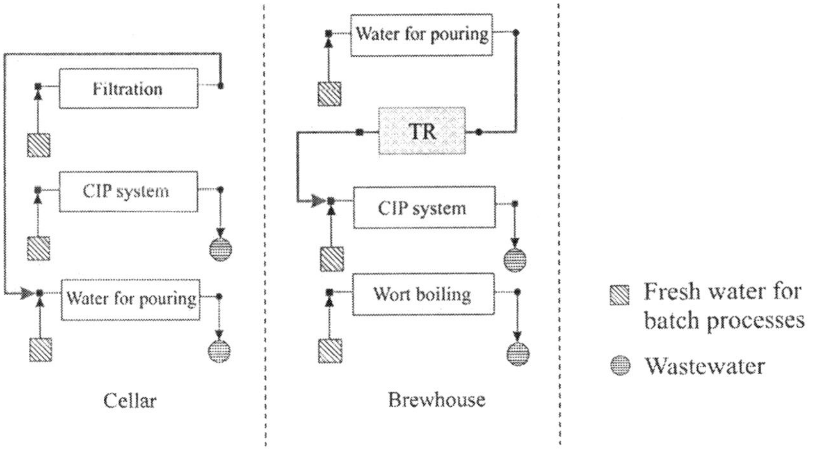

Figure 4 *Water re-use opportunities in a brewery plant*

According to the results of the water network integration in the packaging area,[40] the wastewater from the can rinser can be reused in the pasteurization processes. The freshwater consumption could be reduced by 23 %. The common costs of freshwater and wastewater treatment would decrease by 22 %. No storage tank installation is needed. The net present value of water network reconstruction is positive at 15 % discount rate, and the payback period is 0.29 y (3.5 months)

In the brew house and the cellar the continuous water treatment unit (nanofiltration) was selected for wastewater purification in the optimum water network. Purification enables water re-use from batch material pouring in CIP system. Wastewater from filtration could be directly reused for pouring the batch material. The freshwater consumption could be reduced by 28%. The joint cost of freshwater and wastewater treatment is reduced by 27.9%. Investment costs of the pipeline and the local treatment unit with the storage tank amount to 117,205 €. The required membrane area is 83 m^2. The net present value of the optimal water network is positive at 15% discount rate, and the payback period is 1.3 y.

The price of freshwater has a great influence on the optimal water network. With increased freshwater price, additional re-use and regeneration re-use opportunities are identified, e.g. water re-use between the rinser for nonreturnable bottles and the bottle washer, water re-use between the pasteurizer and the bottle washer, regeneration re-use between wort boiling and the Clean in place, CIP system.

By applying the proposed reuse and regeneration reuse connections, the brewery could save about 25% of the current freshwater demand, and reduce the freshwater cost and wastewater treatment costs by 27%. The volume ratio of water consumption to beer sold in the brewery would decrease from 6.04 L/L to 4.53 L/L, which is important for cleaner production and sustainable development within the company.

4 CONCLUSIONS

Water is used in most process industries for a wide range of applications. Industrial processes and systems using water are today being subjected to increasingly stringent environmental regulations relating to the discharge of effluents and there is a growing demand for fresh water.

The changes taking place and the pace of these changes have increased the need for improved water management and wastewater minimisation. The adoption of water minimisation techniques can effectively reduce overall fresh water demand in water using processes and subsequently reduce the amount of effluent generated, with the result of reducing the costs incurred in the acquisition of fresh water and the costs involved in the treatment of effluent streams.

This field has been developing a rapidly and every year a number of more efficient approaches have been developed. Important methodologies that have been applied in the minimisation of water use and effluent generation in the food and drink industry have been shortly reviewed and demonstrated on selected case studies.

Acknowledgements

Financial support from the EC Marie Curie Chair (EXC) FP6-042618 "INEMAGLOW Integrated Waste to Energy Management to Prevent Global Warming" is gratefully acknowledged.

References

1 A. Y. Hoekstra, *The Value of Water Research Report Series 28*, UNESCO-IHE, 2008.
2 A.Y. Hoekstra and A.K. Chapagain, *Water Resources Management*, 2007, **21**, 35.
3 S. Casani, M. Rouhany and S. Knochel, *Water Research*, 2005, **39**, 1134.
4 D. Dupont and S. Renzetti, *Can. J. Agri. Econ.*, 1988, **46**, 83.
5 A. Fahnich, V. Mavrov and H. Chmiel, *Desalination*, 1998, **119**, 213.
6 Water Footprint Network, www.waterfootprint.org, [16/08/2009]
7 A.K Chapagain and A.Y. Hoekstra, 2004. *Water footprints of nations. Volume 1.* UNESCO-IHE, 2008.
8 J. Klemeš, R. Smith and J.-K. Kim, ed. *Handbook of water and energy management in food processing*, Woodhead Publishing, UK, 2008.
9 R. Smith, *Chemical Process Design and Integration*, John Wiley & Sons, Ltd., UK, 2005.
10 J. Klemeš and S. Perry, *Process Optimisation to Minimise Water Use in Food Industry*, in *Handbook of waste management and co-product recovery in food processing*, ed K Waldron, Woodhead Publising, UK, 2007, 90.
11 B. Linnhoff, D. W. Towsend, D. Boland, G. F. Hewitt, B. E. A. Thomas, A. R. Guy and R. H. Marsland, *A User Guide on Process Integration for the Efficient Use of Energy*, The Institution of Chemical Engineers, Rugby, 1982, last edition 1994.
12 M.M. El-Halwagi, *Pollution Prevention through Process Integration: Systematic Design Tools*, Academic Press, San Diego, 1997.
13 J. Klemeš, V.R. Dhole, K. Raissi, S.J. Perry and L. Puigjaner, *Appl. Therm. Eng.*, 1997, **17**, 993.
14 Y. P. Wang and R. Smith, *Chem. Eng. Sci.*, 1994, **49**, 981.
15 V.R. Dhole, N. Ramchandi, R.A. Tainsh and M. Wasilewski, *Chem. Eng.*, 1996, **103**, 100.

16 N. Hallale, *Adv. Environ. Res.*, 2002, **6**, 377.
17 R. Prakash and U. V. Shenoy, *Chem. Eng. Sci.*, 2005, **60**, 255.
18 M. M. El-Halwagi, F. Gabriel and D. Harell, *Ind. Eng. Chem. Res.*, 2003, **42**, 4319.
19 Z. A. Manan, C. Y. Foo and Y. L. Tan, *AIChE Journal,* 2004, **50**, 3169.
20 S. Bandyopadhyay, M. D. Ghanekar and H. K. Pillai, *Ind. Eng. Chem. Res.*, 2006, **45**, 5287.
21 Y. P. Wang and R. Smith, *Trans. IChemE*, 1995, **73**, 905.
22 D. C. Y, Foo, Z. A. Manan and Y. L. Tan, *J. Cleaner Production*, 2006, **13**,1381.
23 Y. Liu, X. Yuan and Y. Luo, *Chin. J. Chem. Eng.*, 2005, **15**, 369.
24 T. Majozi, C. J. Brouckaert and C. A. Buckley, *J. Env. Management*, 2006, **78**, 317.
25 M. Almató, A. Espuña and L. Puigjaner, *Comp. Chem. Eng.*, 1999, **23**, 1427.
26 B.H. Li and C.T. Chang, *Ind. Eng. Chem. Res.*, 2006, **45**, 5027.
27 A.M. Shoaib, M.A. Said, E.A. Moustafa, D.C.Y. Foo and M.M. El-Halwagi, *Comp. Chem. Eng.*, 2008, **32**, 530.
28 N. Takama, T. Kuriyama, K. Shiroko and T. Umeda, *Comp. Chem. Eng.*, 1980, **4**, 251.
29 C-H. Huang, C-T. Chang, H-C. Ling and C-C. Chang, *Ind. & Eng. Chem. Res.*, 1999, **38**, 2666.
30 A. Alva-Argáez, A. C. Kokossis and R. Smith, *Comp. Chem. Eng.*, 1998, **22**, 741.
31 S. Hul, R. R. Tan, J. Auresenia, T. Fuchino and D. C. Y. Foo, *Proc. Safety Env. Prot.*, 2007, **85**(B6), 1.
32 M. Bagajewicz and M. Savelski, *Trans. IChemE*, 2001, **24**, 600.
33 J-K. Kim and R. Smith, *Proc. Safe. Environ. Prot.*, 2004, **82**, 238.
34 B.H. Li and C.T. Chang, *Ind. Eng. Chem. Res.*, 2006, **45**, 5027.
35 J.F. Gouws and T. Majozi, *Ind. Chem. Eng. Res.*, 2008, **47**, 369.
36 J.F. Gouws and T. Majozi, *Water SA*, 2008, **34**, 291.
37 D. K. S. Ng, D. C. Y. Foo, A. Rabie and M. M. El-Halwagi, *AIChE Journal*, 2008, **54**, 2624.
38 C.L. Chen, C.Y. Chang and J.Y. Lee, *Ind. Eng. Chem. Res.*, 2008, **47**, 7818.
39 C.L. Chen, J.Y. Lee, J.W. Tan and Y.J. Ciou, *Comp. Chem. Eng.*, 2009, **33**, 267.
40 H. Tokos and Z. Novak Pintarič, *J. Cleaner Production*, 2009, **17**, 1465.
41 BREF, *Reference Document on Best Available Techniques in Food, Drink and Milk Industries*, Seville, Spain, 2006.

CHEMICAL AND PHYSIOLOGICAL CHARACTERISATION OF AEROBIC TREATMENT OF RUM DISTILLERY SPENTWASH (DSW) USING *ASPERGILLUS NIGER*

Marie Watson[1], Laurent Corcodel[2], Laurent Dufossé[1] and Thomas Petit[1]

[1] Laboratoire de Chimie des Substances Naturelles et des Sciences des Aliments
Faculté des Sciences et des Technologies, 15 av René Cassin BP 7151, 97715 Saint Denis Messag Cedex 9 France. Email: thomas.petit@univ-reunion.fr
[2] CERF Process Sucrier, BP 315, 97494 Ste Clotilde Cedex, France

1 INTRODUCTION

In Reunion Island, three Rum distilleries share the production of molasses from the local sugar industry to produce Rum. After distillation, 230 000 tons per year of effluent is produced on average. For each litre of alcohol produced, 10 litres of Distillery Spent Water (DSW) are released. These effluents are generally spread on fields but legislation no longer allows the release of them in nature. The relatively low acidity of the effluent (in average pH 4,6) imbalances the soil alkalinity that is favourable to crops. Despite the high potential of fertilization, it has high pollution loads.[1,2] This is related to its dark colour, reducing photosynthesis of the spreaded plants that leads to watercourse eutrophication. The colour of DSW is related to melanoidins present in very large amount in DSW.[3] Pollution is also based on its high Chemical Oxygen Demand (COD). To reduce COD and dark colour, DSW are stored into laguna but this leads to several problems like the increase of insects.[4] DSW can also be treated using anaerobic digestion but the remaining effluent remains difficult to discard because colour and COD remain high.

Several studies have shown that DSW is a potential raw material for industrial use such as the production of bioethanol1, production of Single Cell Protein (SCP),[5,6] production of metabolites like organic acids. In this study, we present our results on physiological and chemical characterisation of the DSW from Reunion Island fermented by the filamentous fungi *Aspergillus niger*. The growth of *A. niger* and the growth characteristics of the cultures were determined. The physical and chemical characteristics of the broth like pH, COD, ashes are also presented.

2 METHOD AND RESULTS

In order to determine the efficiency of the fungus to treat the DSW, cultures using pure DSW were inoculated with 105 spores.mL^{-1} of *A. niger* during 9 days. Because of the high heterogeneity of the culture in DSW media (production of flocks), the growth characteristics were measured for each media using 9 identical flasks containing the same mixture of culture media and cells. Each set of 9 cultures were incubated in the same conditions (room temperature and orbital shaker set at 120 rpm.min^{-1}) before use. Each 24h, cells and supernatant from one culture were separated by filtration to determine biomass and physical and chemical parameters of the fermented effluent. Experiments were reproduced at least in triplicate. DSW1 and DSW2 were obtained from two distilleries of Reunion Island. Unlike DSW2, DSW1 still contains the yeast cells produced during the alcoholic fermentation.

Growth of *A. niger* cells in DSW and in permissive culture medium (Malt medium, Difco) is presented in Figure 1. Surprisingly intensive growth of *A. niger* was observed in pure DSW containing no additional nutriments than the one present naturally in DSW. Biomass content (Dry Weight) was found to be in the same range as in Malt Medium. After a lag phase between 24 and 48 h, exponential growth of cells started and reached a maximum after 150 h of growth. Maximal biomass production was observed for DSW1 (69 $g.L^{-1}$) after 9 days. Growth parameters were also measured for each media. Equivalent growth characteristics were obtained for both DSW1 and Malt media. The generation time and the specific growth rate for DSW1 were respectively tg = 27 h and $\mu = 0.026\ h^{-1}$. The stationary phase was obtained after 144 h of incubation for DSW1. Thus in this DSW1 media, the fungi could find all the necessary nutrients to grow like in a traditional growth medium adapted for fungi. The comparison of the growth of *A. niger* between DSW1 and DSW2 showed that DSW1 gives the best results concerning the production of biomass. Among other possible explanations, this difference is probably a consequence of the residual yeast present in DSW1 which brings to the fungi cells some additional nutriments for growth.

Physical and chemical analyses were performed on supernatant of unfermented DSW1 and on supernatant of DSW1 incubated during 9 days with *A. niger*. Our results were compared to data from literature (largely dominated by publications from India and Brazil) in order to determine the specificity of the local DSW. Results are presented in Table 1. To the exception of pH, all other measured parameters (total solids, suspended solids, ashes, COD and C/N) were always higher in DSW from Reunion than in DSW from other countries.

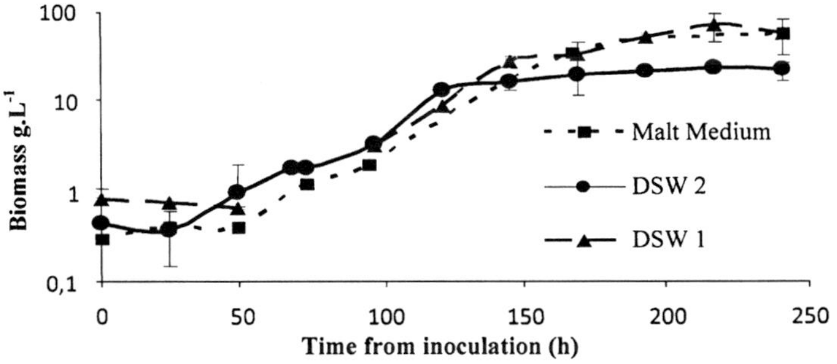

Figure 1 *Growth characteristics of A. niger during 9 days incubation in culture media containing Malt or in DSW from two distilleries from Reunion Island : (■)Malt medium, (▲)DSW 1, (●)DSW 2.*

Table 1 *Physical and chemical parameters of Rum DSW before (unfermented) and after 9 days fermentation with A. niger. DSW1: DSW from Reunion Island.*

Measured parameters	Data from Literature	Unfermented DSW1*	DSW1 after fermentation*	References
pH	4.2 - 5.9	4.6	5.4	**7**
Total solids ($g.L^{-1}$)	4 - 106	189	201	**8, 9**
Ashes ($g.L^{-1}$)	10.7 - 28.9	41.4	43.2	**10**
Suspended Solids (SS) ($g.L^{-1}$)	1.1 - 10.9	12.9	81.9	**8, 11**
COD ($g.L^{-1}$)	32 - 90	146.5	53	**11, 12**
Carbon/Nitrogen ratio (C/N)	7.9	11.8	11.3	**13**

* *Data are average of 3 independent experiments.*

The more important differences were found for Total Solids and COD. DSW1 showed in average values for both parameters about two times higher than published data. These differences are probably the results from differences in sugar cane strains, soils composition and distillation processes. The high content of suspended solids present in unfermented DSW sample (12.92 $g.L^{-1}$) is due to the yeast present during alcohol fermentation.

After 9 days fermentation, very high amount of Suspended Solids were detected (82 $g.L^{-1}$ in average). This value includes filterable biomass produced during the fermentation process. While biomass increased, COD decreased proportionally by means of 64 %. Preliminary results also indicated that Absorbance of DSW measured at 270 nm also strongly decreased during fermentation (from 310 to 145 units).

3 CONCLUSIONS

Analyses of the physical and chemical characteristics of DSW from Reunion Island confirmed that this effluent have a high polluting load. Parameters like COD, SS, and absorbance confirmed this conclusion. Our results clearly show that *Aspergillus niger* can perfectly grow on unsupplemented DSW and we obtained on average about 70 $g.L^{-1}$ of biomass in the better case.

Incubation of *A. niger* in DSW media offers biotreatment potentialities since COD was decreased by a factor 3 during fermentation. According to biomass production, we can also conclude that the fungus has a strong potential for bioproduction. This process still has to be optimized but at this stage of the work, it offers very interesting perspectives for industrial applications. Both aspects, namely bioproduction and bioremediation will be further study with other strains alone or combined in mixed cultures.

References

1 K. Kida, S. Morimura, N. Abe and Y. Sonoda, *Proc. Biochem.* 1995, **30**, 125.
2 A.C. Wilkie, K.J. Riedesel and J.M. Owens, *Biomass and Bioenergy* 2000, **19**, 63.
3 S. Karhaman and O. Yesilada, *Folia Microbiol.*, 2003, **48**, 525.
4 F.J. Fitzgibbon, P. Nigam, D. Singh and R. Marchant, *J. Basic Microbiol.*, 1995, **35**, 293.

5 L. Coulibaly, G. Gourene and S.N. Agathos, *Afr. J. Biotechnol.*, 2003, **2**, 12.
6 S.A. Shojaosadati, R. Khalilzadch, A. Jalilzadeh and H.R. Sanaei, *Resour. Conserv. Recycl.*, 1999, **27**, 125.
7 P.L. Rosalem, S.M. Tauk and M.C. Nunes dos Santos, *Rev Microbiol Sao Paulo,* 1985, **16,** 299.
8 A.M. Azzam and Y.A. Heikel, *J. Environ. Sci. Health, Part A: Environ. Sci. Eng. Toxic Hazard. Subst. Control*, 1989, **24**, 967.
9 G. Cabib, H.J. Silva, A. Giuletti and R. Ertola, *J. Chem. Technol. Biotechnol.*, 1983, **33**, 21.
10 G.J. Sheehan and P.F. Greenfield, "Utilisation, Treatment and Disposal of Distillery Wastewater." *Water Res.*, 1980, **14**, 257.
11 P.R. Lamo and T.J.B. Menezes (1978). "Bioconversao da vinhaca para a producao de biomassa fungíca." *Coletanea do Instituto de Tecnologia de Alimentos* (Brasil), **9**, 281.
12 N.Q. Araújo, A.S. Visconti, H.F. Castro, H.G.B. Silva and M.H.A. Ferraz Filh, "MS Produção de Biomassa Fúngica de Vinhoto." *Brasil Açucareiro* 1976, **88**, 12.
13 V. Parnaudeau, N. Condom, R. Olivier, P. Cazevieille and S. Recous, *Bioresour. Technol.* 2008, **99**, 1553.

DEAD FISH VALORISATION BY ANAEROBIC DIGESTION

A. Esturo, S. Etxebarria, J. Zufía and M. Cebrián

Sustainability of the Food Chain Area, Food Research Division, AZTI-TECNALIA, Astondo Bidea - Edificio 609, 48160 Derio, Bizkaia, Spain

1 INTRODUCTION

Aquaculture in Spain is an activity that is increasing in the last years. According to the figures of the Spanish Ministry of Environment, Agriculture and Fisheries, in 2007 more than 50.000 tons fish were produced, and by 2010, 70.000 tons are expected to be produced. The increase of the aquaculture activity implies also new issues such as the increase of massive mortality episodes. Disposal of big volumes of dead fish becomes a problem for the aquaculture producers, since traditional waste managers (landfill, incinerators) are reluctant to deal with dead fish because of their high content in water and high degradability producing non pleasant odours. Apart of the mentioned difficulty to find a final destiny for massive fish mortality there is a added legal consideration to be taken into account: Regulation EC nº. 1774/2002 regulating the animal by-products not intended for human consumption has to be followed to dispose the dead fish in an appropriate way. According to the Regulation dead fish are classified as by-products category 2, but most of the times, and due to the managing difficulties, these by-products are disposed as category 1, with no possible valorisation options.

Biogas production is one of the most feasible alternatives for organic wastes valorisation, but in the case of fish wastes few studies[1,2,3] have been performed due to their composition (low C/N ratio and high salinity) which is not very adequate for anaerobic digestion.

This study aimed to evaluate the anaerobic digestion process as a valorisation option for fish wastes as energy source with the following two specific objectives: First, to determine the limiting parameters for the anaerobic digestion of this kind of wastes and to establish the starting out conditions of the process. And second, to identify compatible wastes and by-products form other agro-food sources in the surrounding area, for further co-digestion studies to make the process technical and economically feasible.

Experimental trials were carried out to determine the capability of dead fish to produce methane, and evaluate anaerobic digestion as a valorisation alternative.

2 METHOD AND RESULTS

2.1 Materials

Three Spanish aquaculture species, which might cause the most important massive mortality episodes, were characterized: trout, gilthead and sea bass. Physicochemical analysis was carried out on the crushed whole fish to evaluate the initial composition of the waste. Results are shown in Table 1.

Table 1 *Composition of fish wastes of three species referred to wet weight*

Parameter	*Specie*		
	Trout	*Gilthead*	*Sea bass*
Fat content %	7.16	10.8	8.94
Proteins %	18.71	23.41	20.06
Humidity %	69.85	60.81	66.12
Ammonium %	2.26	0.27	0.28
Cd (mg/Kg)	<0.02	0.17	0.08
Cu (mg/Kg)	1.14	0.67	2.25
Cr (mg/Kg)	<0.04	<0.04	<0.04
Ni (mg/Kg)	<0.2	<0.2	<0.2
Pb (mg/Kg)	<0.2	<0.2	<0.2
Zn (mg/Kg)	20.52	22.12	16.39
N-org (mgN/Kg)	4.895	3.339	2.570
NaCl %	2.74	1.30	1.32
Ashes	2.14	2.96	2.41
C/N	6.42	7.58	6.40

The anaerobic digestion pilot trial was done with the trout, which has the higher salinity values (2.74%). Salinity is a restrictive factor in the anaerobic digestion, and the most adverse condition was selected to carry out the trial.

A 42 litres capacity digester was used. Generated biogas was collected in Tedlar bags and analyzed in a gas chromatograph with HWD detector. The trial was done with trout, since it has higher salinity values if compared with other aquaculture species as mentioned above. 4.5 kilo of trout were grinded and diluted with 20 litres of water and introduced into the digester. Volatile Solid (VS9 content was around 95% (dry matter) with a 3% of Total Solids (TS). The inoculums were taken from an urban water treatment plant to be used as starter. The trial was a batch process with mesophylic conditions (37°C) and constant agitation.

2.2 Results

The anaerobic digestion yielded nearly no methane in the beginning. The gas evolution obtained during the digestion is shown in Figure 1.

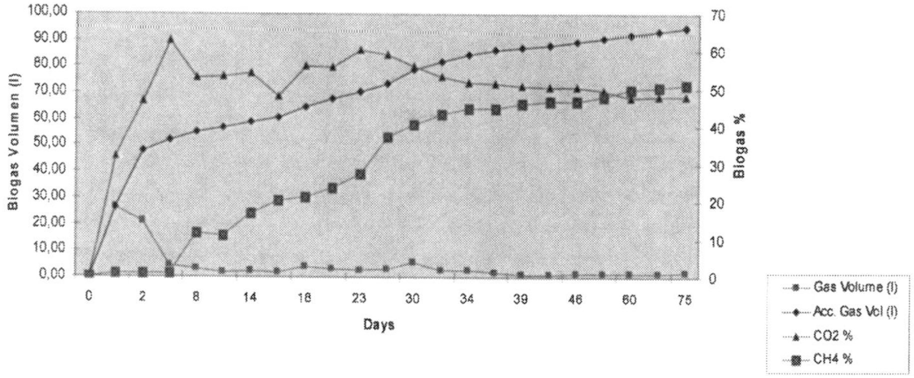

Figure 1 *Evolution of biogas composition and accumulated gas production*

The chromatography results indicate that produced biogas only reaches 50% of methane at the end of the trial, 60 days after the starting out of the reaction. The biogas production is below 5 litres per day and the accumulated gas during the 75 days of the experiment running is 95 litres.

The results indicate a productivity of 11 litre methane per each kilo of dry raw material. If volatile solids are considered, a production of 12 litres of methane per each kilo of substrate can be considered, as shown in Figure 2.

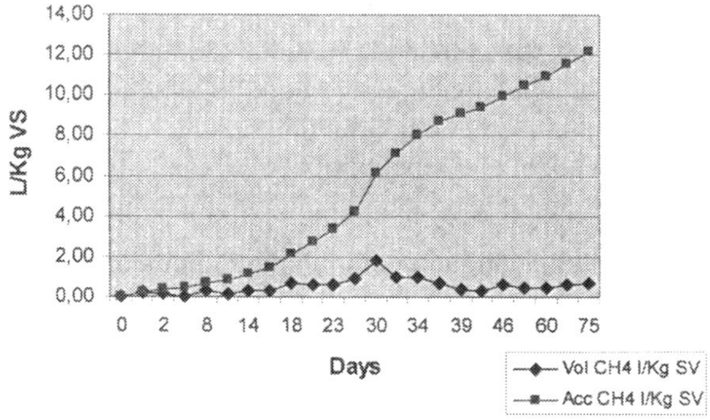

Figure 2 *Evolution of daily and accumulated gas production*

3 CONCLUSIONS

The main results of the experiment indicate that fish wastes are not adequate for anaerobic digestion without a mixture with other co-substrates. The high CO_2 percentage in the produced gas and low methane production indicate that process is unstable, probably due to the inhibitory parameters.

Obtained results give a productivity of about 4% of the maximum expected value (about 360/Kf VS with at least 65% of CH_4). The reasons for this low productivity might be due to the presence of salts in the initial waste (2.54%). Even if later diluted to 0.7%, according to literature,[4] in order to obtain good biogas yields in high salinity substrates a long adaptation time is necessary.

The C/N optimum ratio ranges from 25 to 30. The dead fish analyses indicate that the C/N ratio is 6, meaning an excess of nitrogen, which might limit the process.

The detected high zinc values (3.76 mg/l) after dilution can also contribute to destabilise the process.

The first conclusion is that this substrate is not ideal for anaerobic digestion, but it does generate methane. When co-digested with a proper substrate, it is possible to generate more methane.[5,6] Further research for co-digestion with other agro-food wastes will be performed.

References

1 J.R. Ferguson and M.E. Cape. (1990). Anaerobic digestion of fish wastes. U.S. Patent 4975106.

2 A. Mshandete, A. Kiviasi, M. Rubindamayugi and B. Mattiasson, Anaerobic batch co-digestion of sisal pulp and fish wastes. *Bioresour. Technol.*, 2004, **95**, 19.

3 C. Ward and B. Slater, Anaerobic Digestion of Fish Processing By-products. *Nut. Food Sci.*, 2002, **32**, 51.

4 Y. Chen, J.J. Cheng and K.S. Creamer, Inhibition of anaerobic digestion process: A review. *Bioresour. Technol.*, 2008, **99**, 4044.

5 A.P. Francese, G. Aboagye-Mathiesen, T. Olesen, P.R. Córdoba and F. Siñeriz, Feeding approaches for biogas production from animal wastes and industrial effluents. *World J. Microbiol. Biotechnol.*, 2000, **16**, 147.

6 F.J. Callaghan, D.A.J. Wase , K. Thayanithy and C.F. Forster, Co-digestion of waste organic Solids: Batch Studies. *Bioresour. Technol.*, 1999, **67**, 117.

AGRO-FOOD BYPRODUCTS AND WASTE AS RAW MATERIALS FOR THE TWO-STAGE HYDROGEN FERMENTATION PROCESS

R. Grabarczyk and K. Urbaniec

CERED Centre of Excellence, Warsaw University of Technology, Plock Campus, Jachowicza 2/4, 09-402 Plock, Poland. E-mail: rgrabarczyk@pw.plock.pl

1 INTRODUCTION

The energy policy of the European Union sets up that 20% of energy in 2020 will be produced from renewable resources with the significant role of biomass.[1,2] This implies that in addition to the existing energy technologies based on renewables, new ones should be developed. One of the promising novel technologies is the two-stage hydrogen fermentation process which is the topic of EU-funded research project titled "Non-thermal production of pure hydrogen from biomass – HYVOLUTION".[3]

The two-stage fermentative process consists of thermophilic fermentation and photofermentation. In the thermophilic fermentation, the substrates (simple sugars: glucose, xylose, arabinose etc.) are converted to hydrogen, carbon dioxide and organic (typically, acetic or lactic) acids. The photofermentation consists in reducing organic acids to hydrogen and carbon dioxide. As the organic acids produced in thermophilic fermentation can be used as substrates for photofermentation, both processes can be integrated (Figure 1) offering the possibility of 70% conversion of the substrates to hydrogen and carbon dioxide.[4] The produced hydrogen gas is upgraded in order to obtain required purity.

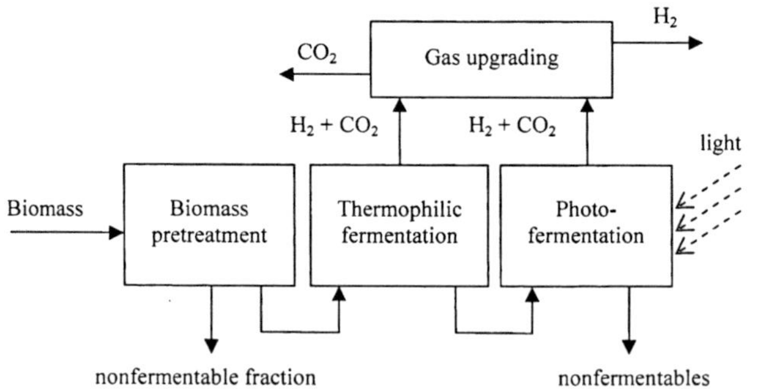

Figure 1 *Scheme of two-stage hydrogen fermentation*

HYVOLUTION project is aimed at developing a blueprint for a complete hydrogen production plant employing the two-stage fermentative process. It is expected that the fermentative process can be applied in decentralised, small hydrogen production units located close to biomass resources.

2 AGRO-FOOD BYPRODUCTS AS RAW MATERIALS FOR FERMENTATIVE HYDROGEN PRODUCTION

A prerequisite for hydrogen fermentation is the availability of feedstock containing simple sugars and therefore the raw biomass must be pretreated using methods that depend on its chemical composition. From this point of view, one should distinguish between three different types of biomass containing:

- simple sugars and disaccharides (sugar beet, sugar cane, sweet sorghum),
- starch (potato, cereals),
- lignocellulose (grass, wood).[5]

Apart from various crops, there are many types of byproducts from agriculture and food industries that can be used for hydrogen production (Table 1). In the face of biofuels vs. food production debate, such raw materials are to be preferred.

Table 1 *Examples of agro-food byproducts*

Type	*Example*	
sugar containing	molasses	
starchy	potato peels	wheat bran[a] wheat middlings[a]
lignocellulosic	beet pulp straw bagasse corn stalks and cobs brewers spent grain distillery spent grain carrot press cake	

[a] contains significant percentage of both starch and cellulose & lignin

For each biomass type, a suitable pretreatment method can be developed on the basis of experience from established industrial processes, e.g. production of sugar, glucose or ethanol. The conversion of starch to glucose can be done by hydrolysis using amylase enzymes in a well known two-step process composed of liquefaction and saccharification.[6] In the case of lignocellulosic feedstocks, fermentable sugars are obtained via hydrolysis of hemicellulose and cellulose. A possible process includes mild acid pretreatment with diluted sulphuric acid, followed by enzymatic hydrolysis.[7] However, cellulase enzymes needed for the hydrolysis of lignocellulose still are too expensive to be used in an industrial scale. Another disadvantage of lignocellulosic feedstocks is the presence of lignin which is an inhibitor of hydrogen fermentation and must be removed in the pretreatment step.

Bearing in mind that the fermentative hydrogen production still is under development, its raw material base can be only roughly estimated. On the basis of European statistical data on the production of selected crops, beer and sugar, the streams of byproducts of agricultural origin were estimated (Table 2).[8] It is reasonable to assume that in the future, some percentage of the harvest of most common crops or byproducts & residues of their processing could be available for large-scale hydrogen production.

Table 2 *EU27 crop production and byproducts streams in 2007*

Crop	Mt/year	*Byproduct*	Mt/year
wheat	120,2	wheat bran[a]	18
		wheat straw[b]	102,2
barley	57,9	brewers spent grain[c]	7,8
		barley straw[b]	49,2
corn grain	47,5	corn cob[d]	23,7
		corn stalk[e]	57
sugar beet	96	sugar beet pulp[f]	25,7
		molasses[g]	4,6
potato	61,4	potato peels[a]	9,21

[a] 15% of wheat, resp. potato production in EU27
[b] 85% of wheat, resp. barley production in EU 27
[c] 20 kg/100 l of produced beer in EU27 (2003 data)
[d] 50% of corn grain production in EU27
[e] 120% of corn grain production in EU27
[f] at 25% dry substance
[g] 4,8% of sugar beet production in EU 27

On the basis of chemical composition of several agro-food byproducts, the attainable yields of two-stage hydrogen fermentation process were calculated (Figure 2). The conversion factors of sub-processes involved were assumed as follows:

- starch hydrolysis - 95%,
- cellulose/hemicellulose hydrolysis - 90%,
- two stage hydrogen fermentation - 70%.

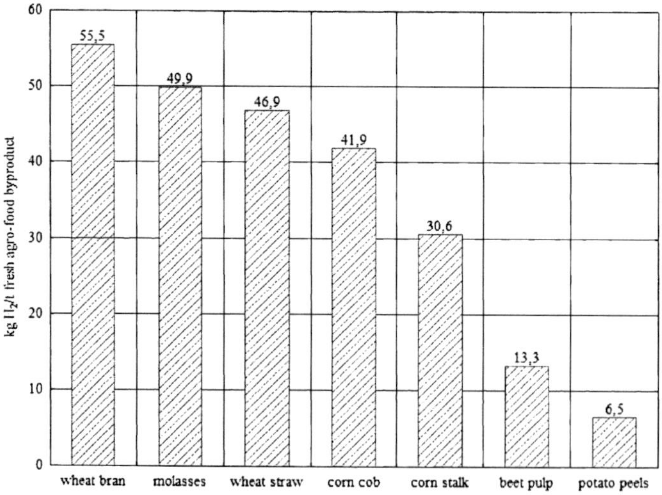

Figure 2 *Attainable hydrogen yields of the two-stage fermentative process*

3 CONCLUSIONS

Hydrogen is recognized as a clean and ecologically friendly energy carrier in the future. There are many types of byproducts from different agro-food industries which can potentially be used as raw materials for fermentative hydrogen production. Sugar containing raw materials (e.g., molasses) are considered for this application in short to medium term, while starchy or lignocellulosic raw materials (e.g., potato peels or wheat straw, barley straw) are taken into account in medium to long term.
In general hydrogen production from agro-food byproducts and residues has a number of advantages, and mainly:

- avoided conflict food vs. energy,
- new market for agro-food industry,
- diversification of energy sources.

Acknowledgements

Support of the European Commission under contract 019825 HYVOLUTION and support in the framework of European Social Fund through Warsaw University of Technology Development Programme, are gratefully acknowledged.

References

1 *Energy For a Changing World. An Energy Policy for Europe - the need for action*, European Commission, Brussels, 2007.
2 *Biomass action plan*, European Commission, Brussels, 2005.
3 P.A.M. Claassen and T. de Vrije, *Int. J. Hydrogen Energy*, 2006, **31**, 1416.
4 T. de Vrije and P.A.M. Claassen, 'Dark hydrogen fermentations' in *Bio-methane and Bio-hydrogen*, eds., J.H. Reith, R.H. Wijffels and H. Barten, Smiet offset, The Hague, 2003, Chapter 5, p 103.
5 K. Urbaniec and R. Grabarczyk, *J. Cleaner Prod.*, 2009, **17**, 959.
6 I.A. Panagiotopoulos, R.R. Bakker, M.A. Budde, T. de Vrije, P.A.M. Claassen and E.G. Koukios, 'Pretreatment of promising biomass feedstocks for biohydrogen production' in *Proceedings of the 16th European Biomass Conference & Exhibition*, Valencia, 2008, p 1670.
7 I.A. Panagiotopoulos and R.R. Bakker, 'Comparative study of different lignocellulosic feedstocks enzymatic hydrolysis for fermentable substrates production' in *Proceedings of the 16th European Biomass Conference & Exhibition*, Valencia, 2008, p 1749.
8 K. Urbaniec and R. Grabarczyk, unpublished results.

Bulk Products for Food, Feed and Non-food Uses

SUSTAINABLE DYES FROM AGRI-FOOD CHAIN CO-PRODUCTS

T. Bechtold[1], A. Mahmud-Ali[1] and S. Komboonchoo[2],

[1]Research Institute of Textile Chemistry and Textile Physics/University Innsbruck, Dornbirn/Austria
[2]Department of Industrial Chemistry, Faculty of Science/Chiang Mai University, Chiang Mai/Thailand

1 INTRODUCTION

The re-introduction of natural dyes into textile dyeing operations is of growing interest due to increasing attention to aspects of water pollution, sustainability of raw materials and products, biodegradability and environmental aspects.[1] Additionally increasing concerns about potential health hazards resulting from the wide use of synthetic organic dyes direct more attention to their possible replacement by colorants obtained from natural resources.

In a long-term research activity the potential of natural colorants in Austria has been reassessed with regard to their use as textile dyes. As a result basic requirements and a general methodology have been worked out to define the position of natural colorants.[2-4]

Traditionally natural colorants were extracted from plants cultivated for this particular use. To keep the production costs for natural colorants competitive, high quality sources for dyes, released from food and agriculture as by-products or wastes, were identified. Representatives are onion peel (flavonoid dyes), pressed grapes (anthocyan dyes), extracted tea (condensed tannins) and bark from timber industry.[5]

In Austria the annual agricultural production of onion is about 110,000 tons, the production of grapes is estimated around 100.000 tons and the cultivation of red beet releases approximately 9.500 tons per year. Wastes from food processing can be estimated with 400 tons red beet peel, 20.000 tons pressed grape residues and 150 tons extracted black tea, released from beverage production ("ice-tea").[6]

There is substantial potential for a re-introduction of natural colorants into textile production, particularly when by-products and wastes from food industry are considered as sources for cheap materials. Figure 1 gives an overview about the various aspects to be considered for a successful re-introduction of natural colorants into technical textile dyeing processes.

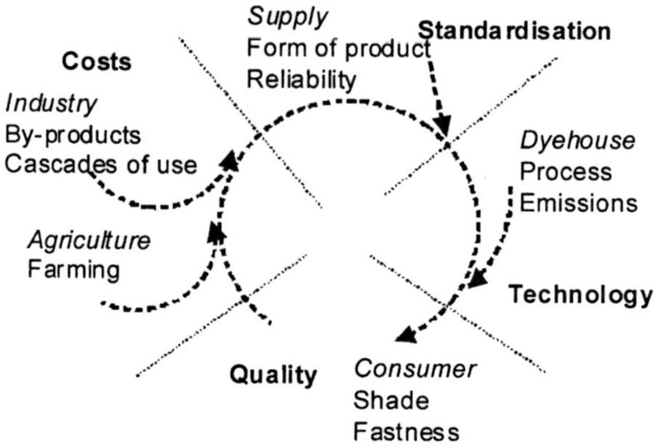

Figure 1 *Process steps and important aspects of natural colorants in technical production*

2 METHOD AND RESULTS

2.1 Definition of Requirements for Natural Dyes

From extensive discussion with farmers, textile dyehouses and market studies, main demands for a re-introduction of natural colorants into state-of-the-art production could be collected.[7] These aspects can be divided into 4 groups:

2.1.1 Farming – Release of Plant Material. To keep production costs low the farming has to be possible with standard equipment without high invests. Low consumption of fertiliser, energy is a prerequisite to keep the overall chemical and energy balances competitive to synthetic dyes. Harvested plant material has to permit simple transportation and storage without cooling or high energy consumption for drying.

2.1.2 Production of a Natural Colorant. For wider application, applicants have to be supplied with a standardised dyestuff product (plant material or extracted colorants) at competitive costs. The dyestuff preparation by extraction has to be considered carefully with regard to energy and chemical consumption, thus plant sources with high extractable dyestuff content are favourable. The use of solvents and chemicals in the extraction step has to be considered with great care, because a high amount of contaminated extracted plant material will be released from the dyestuff extraction step. Extracting dried plant material with water will release approximately three times the mass wet plant material, thus losses of extraction chemicals in this residue will be considerable. Besides the identification of sources which release the required material in appropriate quality and amount, standardisation procedures and suited analytical measures to define the quality of the product have to be provided to the applicant. Plant dyes in many cases consist of a number of chemical compounds, dependent on plant variety, climate and growth and harvesting conditions. Simple standardisation methods without costly analytical equipment have to be developed which enable quality determination and standardisation at the site of production/release of plant material/waste.

A supplier organisation is required which provides the applicant with standardised plant material / extracts and which provides the dyehouse with further supplementary information material e.g. colour charts, fastness properties and recommendations for process conditions. In practice a supply organisation has to be built up which offers similar support as available at present for synthetic dyes.

2.1.3 Dyestuff Application in Technical Processes. According to the present structure of dyehouses, natural dyes should be applicable for all important classes of fibres including cellulose fibres (cotton, flax, regenerated cellulose fibres: viscose, lyocell), protein fibres (wool, hair and silk) and synthetic fibres (e.g. polyamide, polyester, acrylics). The range of available colours should form a complete gamut of dyes covering the full range of basic colours from yellow, red to blue and black.

The dyeing processes have to be applicable on the standard equipment installed in the dyehouses, similar to techniques used at present, extensive modification of technical installation will not be accepted by the dyers. Intermediate storage of big volumes should be avoided. Processes have to be robust, offering at least a minimum reproducibility. Careful calculation of energy and chemical balances for the dyestuff extraction step and the dyeing process have to be done and comparison to the best available technology using synthetic dyes is necessary to demonstrate savings.

Existing limits for polluting chemicals in the waste water reduce the number of applicable mordants considerable. Only iron-salts, alumn and tannin can be used as mordants, other heavy metal mordants such as tin, copper or chromium have to be avoided, with regard to the effluents and heavy metal content on the dyed goods.[7,9]

2.1.4 Consumer Oriented Aspects. A market study to evaluate consumer expectations on natural dyed textiles proved the potential of such products in the market. In any case a critical minimum of fastness properties has to be achieved (e.g. light fastness, wet fastness, wash fastness). Consumer associate natural dyed products with terms such as health, wellness, ecological awareness, responsibility and fairness and are prepared to pay a somewhat higher price for such product.[1] On the other hand, there are concerns with regard to the toxicological potential of plant extracts which could cause allergic reactions. Thus besides technical properties, the selection of plant sources also has to include information about health risks, which favours use of by-products and wastes from agriculture and food industry as these materials are known to exhibit minimum potential for such hazards.

2.2. Onion peel as resource for textile dyes

2.2.1 Plant Material Sources. After harvesting, onions are stored in specialised stores, from where the vegetables are cleaned and packed for delivery to groceries. During the packaging process soil, stem, roots and loose dry peels are removed, damaged onions, small or deformed fruits also have to be removed before packaging. The last step before the onions are packed, is a removal of loose dry peels by vacuum suction.

In our study thus two different fractions could be identified, which were investigated for their potential to serve as a source for the extraction of natural colorants:
Fraction I, containing soil, roots, stem, peels and also non-regular sized onions
Fraction II, released from the vacuum suction process, containing rather clean onion peel.

In Table 1 a gravimetrical analysis of the composition of waste Fraction I is presented.

Table 1 *Gravimetrical analysis of the composition of Fraction I.*

Component	*Mass* g	*Share* %
Peel	429.6	36.0
Stem	194.0	16.3
Soil	83.1	7.0
Onions	478.1	40.1
Dust/Moisture	8.0	ad 100
Total	1192.8	100

The high moisture content in Fraction I facilitated microbial growth, particularly when packing density was rather high. As a result during transportation and storage of the material (Fraction I) temperature inside the bulk increased to 60°C. The microbial processes had to be stopped by unpacking and storage in layers of approximately 30 – 40 cm thickness.

For the extraction of natural colorants from Fraction I, only the peel and stem fractions were studied, while other parts (soil, onions) were removed before extraction. Fraction II contained only onion peel, thus this material could be used for the extraction without further purification.

2.2.2 Extraction. In extraction of onion peel with boiling water quercetin is released as main component. In addition kempferol, quercetin glycosides and tannins have been identified in the extract.[1] Figure 2 shows the chemical structure of quercetin.

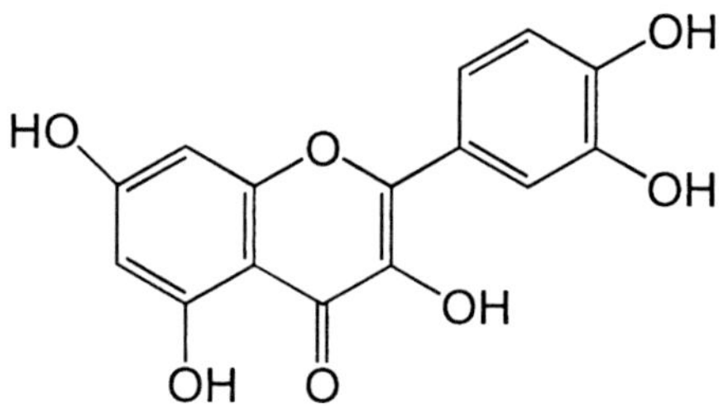

Figure 2 *Structure of quercetin*

For application in textile dyeing operations a standardised dyestuff product would be favourable, however extraction of the plant material and concentration of the extract to a dyestuff product e.g. by re-boiling, will consume considerably high amounts of energy. Thus at present the extraction of the plant material at the site of the use of the hot extract is the recommended procedure, which enables savings of energy due to further use of the hot extract. In this case the dyehouse has to be supplied with standardised plant material, which permits dyeing with acceptable reproducibility. For this purpose simple standardisation methods have been tested with regard to their applicability as standardisation procedures at the site of the plant dye production.

In the present study UV-Vis absorbance of the extract and total phenolics in the extract determined by the Folin Ciocalteau method, were tested as methods to characterise the quality of the onion peel, which regard to the colour obtained in dyeing experiments.

2.2.3 Methods and Analytical Procedures. The aqueous extract was prepared by boiling a certain mass of onion peel with tap water at a mass per volume ratio of 1:20 (g : ml). The extraction was performed for 60 min at the boil. Samples were taken from the extract in regular time intervals.

For photometry, 1 ml of the extract was diluted with 10 ml distilled water and absorbance was measured with a 1 cm cuvette in the wavelength interval between 400 and 700 nm (Zeiss CLH 500/MCS521 UV-vis, Carl Zeiss, Jena, Germany).

Total soluble phenolics (TPH) in the extract were determined with Folin-Ciocalteau reagent according to the method of Slinkard and Singleton using gallic acid as a standard. The extracts were diluted with distilled water to adjust extinction within the range of the calibration curve measured a 765 nm. Results were expressed as mg/l gallic acid equivalents in the extract.[11-13] For extraction and photometrical analysis of phenols two repetitions were performed.

The dyeings were performed by exhaustion method using a liquor ratio of 1:20. For 1 g of goods a dyebath volume of 20 ml was applied. According to a standard procedure the dyeing experiments were performed in beakers at 95°C dyeing temperature. 10 g of bleached wool yarn were used as substrate.

In case of mordanting, $FeSO_4.7H_2O$ or Alum $KAl(SO_4)_2.12H_2O$ were added to the dyebath to give a final dyebath concentration of 5 g L^{-1} $FeSO_4.7H_2O$ (0.016 mol L-1) respectively 5 g L^{-1} (0.011 mol L^{-1}) alum mordant. The detailed dyeing procedure is given in the literature.[8,9]

After dyeing, unfixed dyestuff was removed by three times rinsing with cold water. CIE-Lab values of the dyeings were measured with a tristimulus colorimeter (Minolta Chroma-Meter CR 200, sample diameter 8 mm). Colours are given in CIE-Lab coordinates, L* corresponding to the brightness (100 = white, 0 = black), a* to the red-green coordinate (positive sign = red, negative sign = green) and b* to the yellow-blue coordinate (positive sign = yellow, negative sign = blue).

2.2.4 Extraction of Natural Colorants. The extraction of coloured material and the analysis of total soluble phenolics (TPH) was monitored during standardised extraction of the different plant material fractions. Figure 3 shows the release of colorants measured at the wavelength of maximum absorbance and the TPH in the extract, as function of extraction time for onion peel separated from Fraction I.

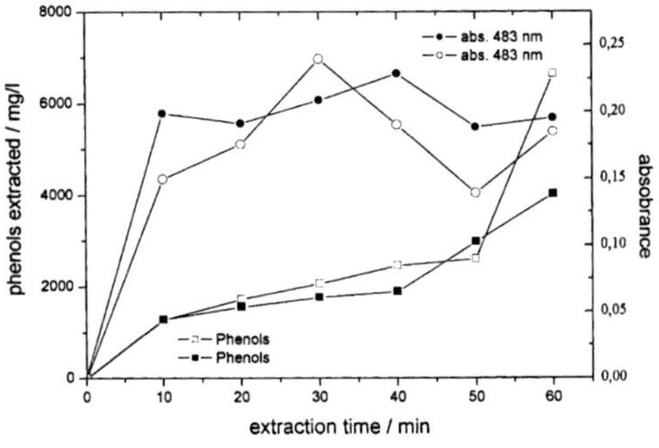

Figure 3 *Release of colorants observed at 483 nm and TPH as function of extraction time*

As can be seen in Figure 3, the major part of coloured components is extracted within 10 – 20 min extraction at the boil. Limited increase in absorbance respectively TPH is observed during extraction between 20 and 40 min. A slight decrease in absorbance is observed above 50 min extraction time, while TPH increases. The increase in TPH can be explained by extraction of other phenolic components from the plant material, which are extractable after a certain period of hot aqueous extraction. Figure 4 shows the absorbance of extracts obtained from different parts of Fraction I (peel, stem in Table 1). In addition a purification of the peels to remove adherent soil by a preceding cold water rinse was studied. The results shown in Figure 4 show the lower dyestuff content in extracts from stem material, which thus is expected to result in lighter dyeings. Extracts from peels, which had been purified immediately before extraction by removal of soil by a cold water rinse, showed highest absorbance.

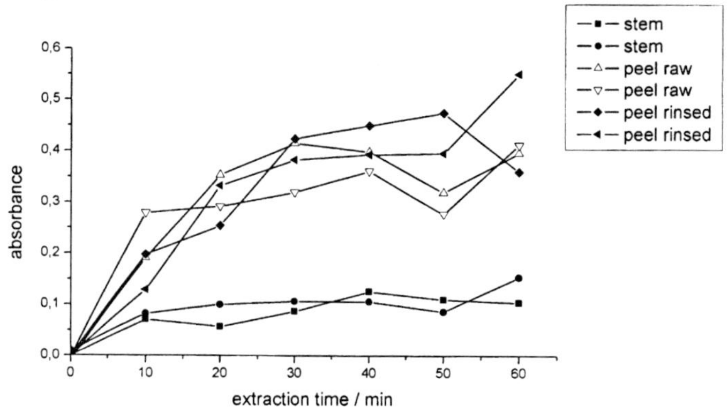

Figure 4 *Absorbance of peel, stem and pre-washed peel from Fraction I material at 483nm*

Figure 5 shows the coupled extraction - dyeing procedure. Ideally the hot plat material extracts are used for the dyeing, which permits substantial savings of thermal energy. After a dyeing time of approximately 15 min at 95°C the solution of the mordant is added into the dyebath. While the traditional pre- or after-mordanting could result in higher efficiency in use of plant extracts, the meta-mordanting procedure has been favoured by modern dyehouses, as intermediate storage of mordant solutions and additional treatment steps can be avoided by the direct addition of mordant into the dyebath.[10] After the dyeing step the dyed material is rinsed three times with cold water.

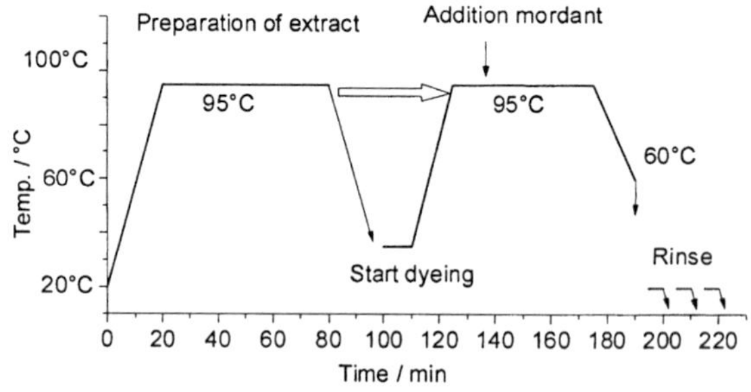

Figure 5 *Optimised extraction – Dyeing procedure*

Table 2 *CIELab Coordinates of dyeings with material from Fraction I on wool.*

Source	*L**	*a**	*b**
Washed peel	30.97	+19.52	+20.93
Washed peel	31.37	+18.79	+20.62
Not washed peel	34.95	+18.38	+23.25
Not washed peel	37.28	+18.78	+24.39
Stem	53.38	+12.72	+29.13
Stem	53.24	+12.14	+29.23

2.2.5 Dyeing Results. The results of laboratory dyeings with extracts from Fraction I material on wool, are shown in Table 2. The CIELab-coordinates correlate with the results of the absorbance measurement given in Figure 4.

As can be seen, the highest colour depth (lowest L*-value) and the most brilliant colour (relatively high colour coordinates a* and b*) is observed with the pre-rinsed onion peel from Fraction I. The adherent soil contaminates the extracts and is supposed to adsorb and precipitate part of the extracted colorant, thus quality of the dyeing obtained with not-washed peel is lower. Extracts from stem yield only light dyeings and thus stems are not useful for natural dye extraction and should be removed during collection. In general the use of Fraction II type material for natural colorant extraction is favourable, because the peels are available already in pure form.

Natural colorants are present in the plant matrix only in relative low concentration, thus after extraction of the dyestuff the major part of plant material is released as wet plant residue. Addition of chemicals or solvents to improve extraction efficiency is prohibited by the fact that high amounts of chemically contaminated waste would have to be disposed, while in case of simple water extraction, the extracted residue can be processes further as usual e.g. composting, animal feed, combustion for energy production.

3 CONCLUSION

While natural colorants have been used since thousands of years, the re-introduction of natural colorants into technical textile dyeing requires the implementation of new concepts. By-products from food and agriculture can serve as cost efficient source for high quality plant material. A supplier organisation is required, which is able to provide applicants of natural colorants with standardised products, supplementary information and technical background.

As an example onion peel are released from farming and packing of onions in huge amounts. Simple analytics e.g. photometry of the aqueous extracts can help to form standardised products with reproducible dyeing behaviour. Also the analysis of TPH (total phenolics) in the extract can provide useful information to optimise extraction time, and asses the quality of a plant source.

The use of water for extraction of the natural colorant permits disposal of the extracted material in the usual processing routes as the material has not been contaminated by chemicals or solvents. This aspect is of significant importance as the amount of processed plant material will be high, as an average 1 kg of plant material will be extracted for 1 kg of dyed goods.

References

1 T. Bechtold and R. Mussak, *Handbook of Natural Colorants*, ISBN 978-0-470-511992, John Wiley & Sons Ltd., Chichester, United Kingdom, 2009.

2 S. Geissler, E. Ganglberger, T. Bechtold, S. Sandberg, O. Schütz, A. Hartl and R. Reiterer, *Potenzial an nachwachsenden Rohstoffen unter den Aspekten der Nachhaltigkeit: Produktion von farbstoffliefernden Pflanzen in Österreich und ihre Nutzung in der Textilindustrie,* Bundesministerium für Verkehr, Innovation und Technologie, Wien, 2001.

3 S. Geissler, E. Ganglberger, T. Bechtold, A. Mahmut, A. Hartl, O. Schütz, *FARB & STOFF - Sustainable Development durch neue Kooperationen und Prozesse, Berichte aus Energie- & Umweltforschung* 25/2003, Bundesministerium für Verkehr, Innovation und Technologie, Wien, 2003.

4 B. Rappl, Ch. Pladerer, M. Meissner, N. Prauhart, G. Roiszer-Bezan, B. Friedrich, E. Egger-Rollig, E. Ganglberger, S. Geissler, Th. Bechtold, R. Mussak, A. Mahmud-Ali, A. Grimm, G. Jasch and E. Freudenthaler, *TRADEMARKFarb&Stoff. Von der Idee zum marktfähigen Handelsprodukt: Pflanzenfarben für die Textilindustrie, Berichte aus Energie- & Umweltforschung 21/2006,* Bundesministerium für Verkehr, Innovation und Technologie, Wien, 2006.

5 T. Bechtold, P. Kaulfuss, A. Mahmud, S. Geissler and E. Ganglberger, *Melliand Textilber.*, 2004, **85**, 268.

6 T. Bechtold, A. Mahmud-Ali and R. Mussak in *Handbook of waste management and co-product recovery in food processing*, ed. K. Waldron, Woodhead Publishing Ltd, Cambridge, England, 2007, p502.

7 T. Bechtold, A. Mahmud-Ali, E. Ganglberger and S. Geissler, *Int. J. Environ. Waste Manage.*, 2008, **2**/3, 215.

8 T. Bechtold, A. Mahmud-Ali and R. Mussak, *Dyes Pigm.*, 2007, **75**, 287.

9 T. Bechtold, A. Turcanu, E. Ganglberger and S. Geissler, *J. Cleaner Prod.*, 2003, **11**, 499.

10 H. Schweppe, *Handbuch der Naturfarbstoffe*, ISBN 3-933203-46-5, ecomed Verlagsgesellschaft, Landsberg/Lech, Germany, 1993.

11 A. L. Waterhouse, *Determination of Total Phenolics*, in *Current Protocols in food Analytical Chemistry*, I1.1.1-I1.1.8, Wrolstad, R.E., Wiley, 2001.

12 V. L. Singleton, R. Orthofer and R. M. Lamuela-Raventos, *Methods Enzymol.*, 1999, **299**, 152.

13 K. Slinkard and V. L. Singleton, *Am. J. Enol. Vitic.* 1977, **28**, 49.

FOOD FRACTION VALORISATION AS ANIMAL FEED IN THE BASQUE COUNTRY

S. Ramos, A. Esturo, S. Etxebarria and J. Zufia

Sustainability of the Food Chain Area, Food Research Division, AZTI-TECNALIA, Astondo Bidea-Edificio 609, 48160 Derio, Bizkaia, Spain

1 INTRODUCTION

New waste management policies aim to reduce the volume of waste going to dumps and incineration plants, encouraging at the same time the valorisation. It's due to a dumps overexploitation and the low quantities of recycling and valorisation.

In this context, the environmental Framework Plan of the Basque Country has defined some specific goals for 2012. On the one hand, a specific objective for 2012 is to reduce until 45% the urban wastes sent to dumps. On the other hand, to reduce in 50% the biodegradable wastes sent to dumps for 2009, by means of a responsible management of the natural resources.

2 OBJECTIVES

This work studies the sanitary, technical and economical viability and evaluates the potential revalorisation of wastes produced by the delivery and commercialization of fresh foods, to be used later as raw materials for animal feed production.

The specific objectives of the project are:

To get a reduction of the volume of biodegradable wastes generated by food markets and hypermarkets

To provide new raw material sources to animal feed producers

3 MATERIALS & METHODS

The study was carried out in the Autonomous Community of the Basque Country (Spain). The food fractions were taken from markets and dumps, identified in origin and later characterized. These wastes were separated in seven fractions: vegetables, fruits, meats, fish, shellfish, eggs and diaries. The food fractions were classified according to their potential valorisation as animal feed.[1, 2]

Table 1 *Potential valorisation options for each food fraction*

Valorisation options	*VEGETAL*		*ANIMAL*				
	Fruits	*Vegetables*	*Meat*	*Fish*	*Shellfish*	*Dairy*	*Eggs*
Direct food for animals: Ruminants, Pigs, Fowl, Fish	X	X			X	X	X
Direct Food for animals: Pets	X	X	X	X	X	X	X
Direct food for animals: Zoo animals, circus, reptiles, dog, pounds, worms	X	X	X	X	X	X	X
Pet food producers	X	X	X	X	X	X	X
Chewable pet accessories	X	X	X	X	X	X	X

Samples were taken from 10 supermarkets and analysed in laboratories to evaluate the hygienic requirement.[3] In addition, sanitary, technical, and market requirements were considered to evaluate the real viability of these fractions.

Finally results were evaluated by an expert group to identify the real issues in the valorisation as well as to propose solutions to cope with the difficulties.

4 RESULTS & DISCUSSION

4.1 Food fraction valorisation

In the autonomous Community of the Basque Country, there are 2 urban solid treatment plants and 6 dumps. They process about 15.000 ton of urban solid wastes coming from markets, hypermarkets and household wastes, which are produced as shown in Table 2:

Table 2 *Annual food waste expected in waste treatment plants*

Fraction	*Quantity (%)*
Vegetables	27
Fruits	28
Meat	4
Fish	13
Shellfish	7
Others	20

Most of the retailers and markets perform the waste management with waste separation at source for further recycling (plastics, wood, glass), but some dispose the food wastes in wooden or plastic boxes, which is an inconvenience to process the food wastes. Within the food wastes, vegetables and fish wastes are more common, being meat, dairy and bread wastes very scarce.

4.2 Sanitary viability

After the analysis of the applicable regulations,[4,5] EC regulation 1774/2002 and Spanish Royal Decree 56/2002, the following groups of parameters were analysed:

Table 3 *Parameters of interest*

Group of parameters	*Acceptable legal levels*
Nutritional	X
Heavy metals	X
Microbiology	-
Mycotoxins	X
Pesticides	-
Chemical compounds	X
Antinutritious factors	X
Others	X

Studied food fractions fulfilled the hygienic and sanitary requirements to be used as animal feed, except some samples that gave high values in microbiology and pesticide levels. This fact was due to incorrect manipulation. If product is properly handled, acceptable levels can be achieved.

4.3 Technical and market viability

There is a real demand of raw material for animal feed production (300.000 tons/year consumption for feed raw materials, 21 authorized plans for animal feed production, and 2 plants for fish and meat meal production). However, there are some acceptance criteria to be fulfilled and technical difficulties to launch these by-products (Table 4).

Table 4 *Technical and market requirements*

Acceptance criteria for raw materials[1]	*Technical difficulties to launch these by-products*
Vegetables by-products dehydrated to 12%	Logistics and transport costs
Blended	Assure the traceability
High nutritional value	Each representative of the valorisation chain has to get the appropriate license to operate
Vegetable and animal separation	Low protein concentration in vegetable by-products
No separation need among vegetables	Storage without humidity (better quality by-products)
Previous selection of by-products to be valorised (rotten, rejected products)	Overcome the reluctance of stockbreeding group to new products
Free of packaging	

[1] Coming from feed, fish meal and oil fish producers

To overcome the mentioned issues a logistic plan was proposed (Table 5).

4.4 Action Plan

This action plan gave rise to 3 specific plans in order to reuse separately fish, vegetable and meat fractions. These plans were calculated for the 10 most important fresh markets and hypermarkets of the Basque Country.

Table 5 *Proposed logistics for collection and valorisation*

STORAGE	Closed containers for collection by a truck
TRANSPORT	Collection trucks will transport food fractions separately until a half way plant for vegetables and straight for fishmeal (Bermeo)or meat meal (Bilbao)
HALF WAY PLANT for vegetables	Separation and selection of quality vegetable fractions Dehydration to 12% to be finally delivered as pellets (in bulk, big bags, etc) Proposed location in Eibar (smaller effort in transport)
PLANT	Animal origin fractions (meat and fish) transported to authorised plants Delivered 100% free of packaging
	Rejected fractions to authorised disposal plants (incineration, compost, dumps)

Besides this proposed logistics, the following were also defined:

1. The required infrastructure for storage, transport and logistics, transformation of food fractions, and delivery to final valorising agent
2. Necessary legal proceedings for food fractions producers, carriers, processing agent and final valorising representative
3. Hygienic handling requirements
4. Labelling and traceability control of the product
5. Investment and operating costs, supposing we were to run this valorisation business.

5 CONCLUSIONS

Firstly, food fractions can be used as a new resource of raw materials for feed, if properly handled to avoid sanitary and hygienic issues, but an adequate separation at origin is crucial. Besides this, valorisation is possible when a responsible is identified in each step and if the traceability of the product is guaranteed at any point.

Secondly, the confidence of feed users on food coming from markets and hypermarkets has to be improved.

Finally, assuming a no profit no loss business, collection cost turn to be far lower than current's ones.

References

1 Ministerio de Agricultura Pesca y Alimentación. 2007. Libro blanco Subproductos de Origen Animal no Destinados a Consumo Humano. Madrid, 392 pp

2 García, A.J., Esteban, M.B., Márquez, M.C. and P. Ramos. *Biodegradable municipal solid waste: Characterization and potential use as animal feedstuffs.* Waste Management, 2005, **25**, 780-787.

3 Norma UNE-EN ISO 6497:2002. Junio de 2006. Alimentos para animales. Toma de muestras
4 Reglamento (CE) n° 1774/2002 del Parlamento Europeo y del Consejo de 3 de octubre de 2002 por el que se establecen las normas sanitarias aplicables a los subproductos animales no destinados al consumo humano. Diario Oficial n° L 273 de 10/10/2002 p. 0001 – 0095
5 Real Decreto 56/2002, de 18 de enero, por el que se regulan la circulación y utilización de materias primas para la alimentación animal y la circulación de piensos compuestos.

ASPARAGUS FIBRES AS REINFORCING MATERIALS FOR DEVELOPING 100% BIODEGRADABLE PACKAGING

S. Jaramillo-Carmona,[1] R. Guillén,[1] C. Escrig-Rondan,[2] J.M. Fuentes-Alventosa,[1] G. Rodríguez,[1] A. Lama[1], A. Jiménez-Araujo,[1] J. Fernández-Bolaños,[1] and R. Rodríguez-Arcos[1]

[1] Departamento de Biotecnología de Alimentos. Instituto de la Grasa (CSIC). Avenida Padre García Tejero, 4, 41012 Sevilla, Spain
[2] AIMPLAS Instituto Tecnológico del Plástico. València Parc Tecnològic. C/ Gustave Eiffel, 4. ES-46980 PATERNA (Valencia), Spain

1 INTRODUCTION

Biodegradable plastic materials have grown to be an alternative to traditional plastics, mainly in those applications with a short-medium shelf-life whose elimination represents environmental and economical problems.[1] Many biodegradable materials have been developed during the last years, but one of its main limitations consists on its poor resistance of continuous medium-high strengths. In order to increase its mechanical resistance, these materials can be modified by the addition of reinforcing fibres from natural origin. So, the mechanical properties are improved without loosing the total biodegradability of the final product.[2]

Asparagus fibre from the industrial processing is an important by-product that contains about 60% cellulose and 30% hemicelluloses, mainly arabinoxylans. This material can not be used as reinforcing in its original form. However, our Group has developed a steam-explosion system that allows to obtaining modified asparagus fibres with high reinforcing capacity. The hydrothermal treatment of asparagus by-products, in addition to releasing several compounds of industrial interest, such as sugars and polysaccharides; generates a solid residue that consists on a "modified" fibre, rich in cellulose and formed by individual fibres, which is suitable as reinforcing material for designing biodegradable plastics.[3]

The main objective of the present work consists on re-valorisating a vegetable waste by the development of biodegradable, thermo-conformed and injectable materials with high mechanical resistance from asparagus fibres that are previously modified by thermal and chemical procedures. The applications of the new materials are focussed on packaging (thermoconformed pieces), agricultural (thermoconformed and injectable materials) and pirotechnic sectors (injected pieces).

2 METHODS

2.1 Preparation of modified fibres from asparagus by-product

By-product generated during the processing of green asparagus was used for the present work. This consists on the final portions of the stems that are discarded before asparagus canning and represents about 50% of asparagus spears. The asparagus by-product, which was obtained from Centro Sur S.C.A. (Huétor–Tajar, Granada, Spain), was sent to our laboratory within 24 h and held at 4°C prior to its processing.

Raw fibre from asparagus by-product was modified by using a steam-explosion process previously developed by our Research Group. Hydrothermal treatment of the samples was carried out in a flash hydrolysis laboratory pilot unit designed at the Instituto de la Grasa (Seville, Spain). After hydrothermal treatment, the solid and liquid products were recovered and separated by filtration through filter paper using a Büchner funnel. The insoluble residue, which constitutes the treated asparagus fibres, was dried, after washing with acetone, and stored for further treatment.

2.2 Compounding, Co-extrusion and Thermo-conforming

The fibres obtained under the above described conditions (treated fibres) were added to biodegradable thermoplastics, such as Mater-Bi and Bionelle, in order to developing new materials with improved mechanical properties. Mixtures consisting on biopolymer + asparagus treated fibres were used for obtaining sheets of the new biodegradable material, by using conventional sheet co-extrusion equipment. Treated fibres and thermoplastic matrix were mixed in a co-rotating twin-screw extruder, the temperature was 170-175°C, operating pressure 2.7-3.1 MPa, and extrusion rate 200 rpm. Mono-layer and multi-layer films were obtained by following the process above explained. The material recovered from the mixer was compression-molded with a MECAMAQ PHF 50 instrument (Milano, Italy). The temperature was fixed at 150°C, operating pressure was 0.2-7.5 MPa and time of compression-molded was 6-12 min.

2.3 Mechanical characterisation

Mechanical properties of the biodegradable sheets were determined by two different methods. Traction and flexion tests were carried out in a universal testing machine (INSTRON 4202). The impact assays were made by using an instrumented pendulum provided of a 15 J hammer (CEAST). Flexure properties were determined according to UNE-EN ISO 178: 2003. The Charpy-impact strength was calculated according to UNE-EN ISO 179-1.

2.4 Visual appearance

Sheets from the new biodegradable material, consisting on biopolymer + 10% asparagus fibre were analysed for their optical properties. Colour, gloss, transparency and smoothness from mono-layer and multi-layer films were compared.

3 RESULTS

3.1 Preparation of modified fibres from asparagus by-product

Asparagus fibre from asparagus processing is a by-product that represent about 50% of total asparagus spear. This fibre consists on 60% cellulose and 30% hemicellulose, mainly arabinoxylans. These raw fibres are not suitable for being used as reinforcing material for developing new biodegradable films, due its poor relation form, high water content and insufficient cellulose content. In order to improve the physico-chemical properties of asparagus fibres, we have applied a treatment called "Steam Explosion" in which high temperatures (160-240 °C) and pressures (5-40 atm) are applied for short periods of time (2-10 minutes). Under these conditions, there is a depolymerisation of polysaccharides (mainly hemicelluloses) and a breakage of the lignin-carbohydrates bonds, resulting in the solubilisation of lignin fragments of low molecular weight. The modified fibres obtained by using the above described process, presented greater strength capacity and thermal stability than the original raw asparagus fibres. These improved characteristics are related to the higher relation-form found in the treated fibres, as well as their lower water content and greater cellulose content.

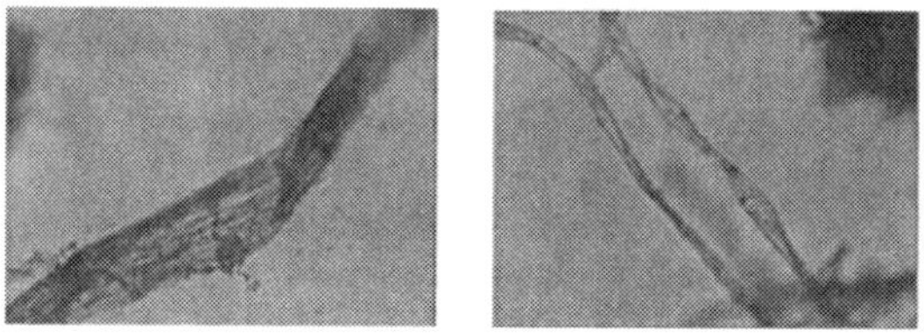

Figure 1 *Asparagus fibres microscopy: raw fibres (left) and steam-exploded fibres (right)*

3.2 Compounding, Co-extrusion and Thermo-conforming

Composites were prepared by adding treated asparagus fibres to biopolymers and mixing them in a co-rotating twin-screw extruder. Then, monolayer and multilayer sheets of the new biodegradable film were obtained by extrusion and co-extrusion process respectively. Multilayer films comprised an internal layer of Mater-Bi N + asparagus modified fibre and two external layers from Mater-Bi N (size relation: 10 / 760 / 10 μm). These external layers, despite being very thin, provided a protective coat that improved the physico-chemical properties of the new biodegradable films.

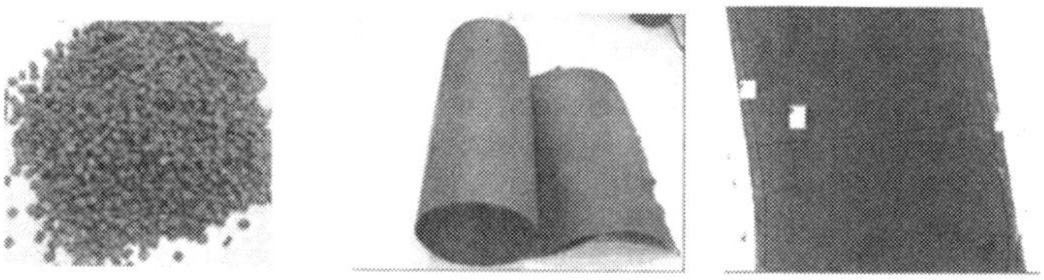

Figure 2 *Composites, monolayer and multilayer sheets from the new biodegradable film*

This material was used for designing thermo-conformed pieces by using a MECAMAQ PHF 50 instrument. The pieces from the multilayer films showed better thermo-conforming properties, including mechanical and optical characteristics.

Figure 3 *Thermo-conforming pattern (left) and thermo-conformed piece (right) from the new bio-film*

3.3 Mechanical characterisation

The mechanical properties of the new bio-film were determined by traction and flexion assays. Several parameters related to elasticity, resistance and strength properties were evaluated from both monolayer and multilayer sheets of the new material, and the results were compared to those from the biopolymer without addition of asparagus fibres.

Table 1 summarises the values of flexion parameters, expressed as Flexion modulus, Force maxima and Arrow to Force maxima. As it can be observed, the addition of asparagus fibres meant an increase of the forces required for deforming the sheets of the film, whereas there were no significant differences for the elongation values measured from both types of films (with and without addition of asparagus fibres). It can be concluded that asparagus fibres increased the elasticity of the new bio-materials described in the present work.

Table 1 *Flexion mechanical parameters from biodegradable materials.*

Material	Flexion modulus (MPa)	Force maxima (MPa)	Arrow to F. max (mm)
Mater-Bi N	202.1 ± 5.8	9.00 ± 0.14	14.01 ± 0.20
Mater-Bi N + asparagus fibre	415.4 ± 21.1	14.29± 0.46	13.24 ± 0.62

The resistance or strength of the bio-films was also tested. For this purpose, it was employed an impact assay, consisting on applying a force maxima of 4 J on the sheets of the bio-films and calculate the kinetic energy required for breaking the film sample.

The values of the impact parameters from the bio-films are shown in Table 2.

Table 2 *Impact values from biodegradable materials.*

Material	Impact (MPa)	s (MPa)
Mater-Bi N	41	4
Mater-Bi N + asparagus fibre	45	3

From all these results it can be concluded that the new biomaterial containing asparagus fibre is more rigid, as it presents a greater flexion modulus, without loosing flexibility, since the arrow and resistance to impact are kept.

On the other hand, a slight increase of impact values means that the asparagus fibre acts as a good reinforcing material. Resistance to impact is a very sensitive mechanical property, closely related to the quality of the interface, which, in this case, consists on asparagus treated fibre.

Finally, the comparison of the monolayer and multilayer sheets from the new biodegradable material revealed that their rigidity was similar, but the elasticity of the second was up to 400% higher that that form the monolayer sheets. The results from the mechanical properties of monolayer and multilayer films are shown in Table 3.

The analysis of the results below show evidence that the addition of very thin external coats on the central layer of Mater-Bi + asparagus fibre resulted in a great improvement of the mechanical characteristics of the new biodegradable film, mainly those related to its elastic properties.

Table 3 *Mechanical characteristics of mono- and multilayer sheets of the new biomaterial.*

New biofilm (Mater-Bi + asp. fibre)	Strength max (MPa)	% Force max	Strength to break (MPa)	% to break
Monolayer sheet	9.51 ± 0.69	6.86 ± 0.27	8.93 ± 0.65	7.78 ± 0.32
Multilayer sheet	10.52 ± 0.56	12.62 ± 0.19	7.90 ± 0.18	32.11 ± 4.24

2.4 Visual appearance

Thermo-conformed pieces from monolayer and multilayer sheets of the new bio-film were compared for their visual appearance. Both films were suitable for thermo-conforming, but several differences were found between mono- and multi-layer films. So, the first had a heterogeneous distribution of asparagus fibres, whereas the second were much more homogeneous and had better physico-chemical properties, such as colour, brightness and smoothness, as it can be observed in Figure 4.

Figure 4 *Sheets from the new biodegradable film prepared in monolayer (left) and multilayer form (right)*

4 CONCLUSION

Asparagus fibres resulted in being adequate as reinforcing material for the development of new biodegradable materials, consisting on mixtures of biopolymers (Mater-Bi N) and "steam-exploded asparagus fibres". Multilayer co-extruded sheets from the new bio-film presented better thermo-conforming properties than the mono-layer sheets. Main improvement consisted on gaining elongation capacity, which makes easier the transformation of the bio-films in different packaging forms.

References

1 B. Kapoor and M. Bhattacharya, *Polym. Eng. Sci.*, 1999, **39**, 676.
2 O. Odusanya, U. Ishiaku, B. Azemi, B. Manan and H. Kammer, *Polym. Eng. Sci.*, 2000, **40**, 1298.
3 M. Anglès, F. Ferrando, X. Farriol and J. Salvadó, *Biomass Bioenergy*, 2001, **21,** 211.

INCREASING PROTEIN EXTRACTION YIELD FROM DUCKWEED (*Lemna obscura*) WITH AN AMMONIA TREATMENT

L. Urribarrí,[1] J. Ríos[1] and A. Ferrer[2]

[1] Analytical Instrumentation Laboratory. Chemistry Department, Science Experimental Faculty. University of Zulia. Maracaibo, Venezuela.
[2] Zulian Institute of Technological Research. La Cañada de Urdaneta. Venezuela. e-mail: alexis.ferrer@gmail.com

1 INTRODUCTION

There is a great protein deficit both for human and animal feeding. In countries like Venezuela, there are no appropriate climate conditions to grow cereal grains and legumes. Therefore, new protein sources are needed.

Recently, there has been a massive growth of duckweed (Lemna obscura),[1] an aquatic plant, at the Maracaibo Lake, covering between 4 and 9% of the lake surface. When it is young, its protein content may be as high as 45% and fiber content is around 10-15%. However, when the duckweed has reached the shore, its protein content usually decreases to 6-25% and the fiber content increases up to 40%. Unfortunately, a great part of the protein is chemically associated with the cell wall. As a result, the protein has a very low digestibility.[2]

Ammonia has proved to be effective to separate vegetable proteins from the fibers. In this way, a protein concentrate may be prepared with a much higher nutritional value.[3] Protein extraction yield from untreated duckweed is low.[2]

The objective of this research was to develop an efficient method of cytoplasmic protein extraction from ammonia-treated duckweed.

2 METHOD AND RESULTS

2.1 Material and Ammonia Treatment

Lemna obscura, collected from the Maracaibo Lake (Venezuela) was used. The duckweed was dried in a force convection oven at 55°C for 72 h and milled to a 2 mm particle size. A representative sample of 300 g of duckweed was processed in a 2 kg biomass ammonia pilot plant (TD). The processing conditions were: 0.5 kg ammonia/kg dry matter, 60% moisture content (d.w.b.), 75°C for 2 min. Untreated duckweed was used as a control (UD).

2.2 Chemical composition of untreated and ammonia-treated duckweed

Moisture content,[4] crude protein[5] and cellulose, hemicellulose and lignin[6] were determined in the raw and ammonia-treated materials. Table 1 shows the composition and it can be seen that the duckweed is in a mature state, since protein content is relatively low (13.23%) whereas neutral detergent fiber is high (55.61%). It looks more as a forage that as an aquatic plant. The ammonia treatment tended to decrease the hemicellulose content by partial solubilization and increased solubles as it has occurred in other lignocellulosic materials.[7] Cellulose and lignin did not change ($p>0.05$).

Table 1 *Chemical composition of untreated and ammonia-treated duckweed.*

Samples	*S*[a] (%)	*H*[a] (%)	*C*[a] (%)	*L*[a] (%)	*Ash*[a] (%)	*CP*[a] (%)	*Moisture* (%)	*NDF*[a] (%)
UD	44.39a	22.98b	17.25a	2.46a	12.97a	13.23a	9.79	55.61a
TD	48.22b	21.67a	17.52a	3.24a	9.35b	13.84a	12.72	51.78b

UD: untreated duckweed. TD: ammonia-treated duckweed. S: Solubles. H: Hemicellulose. C: Cellulose. L: Lignin. CP: Crude protein. NDF: Neutral detergent fiber. [a] Dry matter basis. Different letters mean significant differences. α: 0.05.

2.3 Protein Extraction

White protein extraction from untreated duckweed was carried out with distilled water (pH 5.79) at 45°C, 1:15 solid/liquid ratio for 45 min[2] and turned out to be 25.54%.

For ammonia-treated duckweed, distilled water and a calcium hydroxide solution were used as extracting agents. All the extractions were carried out in triplicate with 5 g of dry matter with constant stirring, varying time (10, 20, 30, 45 and 60 min), pH (calcium hydroxide solutions at pH 10, 11, 12 and 12.60, and distilled water pH 5.79), solid/liquid ratio (1:10, 1:15) and temperature (30, 45, 60, 75 and 90 °C). Extracts were refrigerated to eliminate green proteins by precipitation. After centrifugation, the white protein present in the extracts (true protein) was determined by the Lowry's modified method.[8] Protein extraction yield was determined with respect to initial crude protein content.

The optimal pH for the protein extraction from ammonia-treated duckweed was determined at 30°C, 1:15 solid/liquid ratio for 45 min. Distilled water (pH 5.79) and calcium hydroxide solutions (pH 10 to 13) were used as extractants. Table 2 presents the results.

Table 2 *Optimal pH determination.*

Extractant pH	*Extraction yield (%)*
5.79	27.45b
10.00	26.99b
11.00	26.97b
12.00	27.51b
13.00	22.33a

Different letters mean significant differences. α: 0.05.

There is no need to increase pH to improve the protein extraction yield from duckweed. It can be carried out with water. For a pH 13, hydrolysis of proteins begins to occur. On the contrary, it is necessary to increase the pH in grasses.[3]

The optimal time for the protein extraction from ammonia-treated duckweed was determined at 30°C and 1:15 solid/liquid ratio with distilled water. Table 3 presents the results.

Table 3 *Optimal time determination.*

Extraction time (min)	*Extraction yield (%)*
10	28.98b
20	31.29c
30	28.39b
45	27.45b
60	24.59a

Different letters mean significant differences. α: 0.05.

The protein yield did not increase over 20 min. Therefore a 20 min time was chosen for the rest of the extractions. Increasing the time shows an opposite effect, likely due to insolubilization of proteins by denaturation.

The optimal temperature and adequate solid:liquid ratio were determined with distilled water for 20 min (Figure 1).

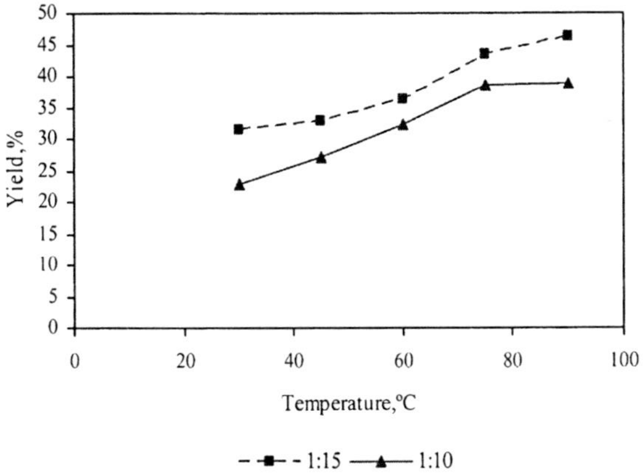

Figure 1 *Determination of optimal temperature and solid/liquid ratio*

Protein extraction yield significantly ($p<0.05$) increased as temperature increased and the solid/liquid ratio decreased. The best conditions were 90°C and 1:15 solid/liquid ratio with a 46.41% yield. It is not convenient to use a temperature higher than 90°C to avoid protein degradation.

Extraction yields at optimal conditions for untreated and ammonia-treated duckweed are compared in Figure 2.

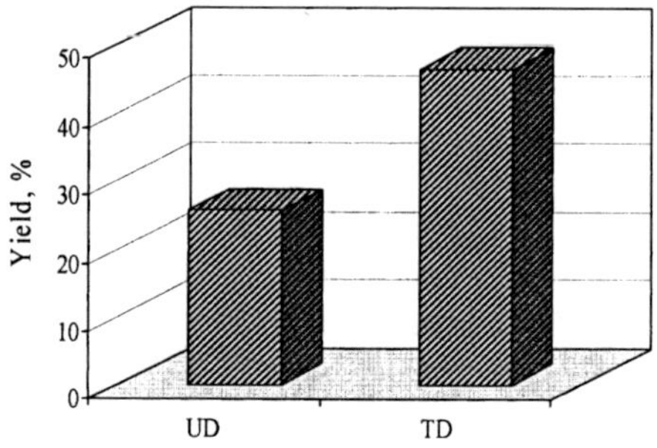

Figure 2 *Effect of the ammonia treatment on protein extraction yield at optimal extraction conditions. UD: untreated duckweed. TD: ammonia-treated duckweed*

The ammonia treatment considerable increased the protein extraction yield (25.54 vs. 46.41%) as it has been found in grasses[3] and cassava leaves.[9] Ammonia treatments opens the lignocellulosic matrix mainly by breaking ester linkages between polymers such as hemicellulose, lignin and proteins. As a result, proteins are more accessible.

3 CONCLUSIONS

The ammonia treatment increased (1.82-fold) the white protein extraction yield from duckweed.

References

1 E. Landolt and R. Kandeler, *Biosystematic investigations in the family of duckweeds (Lemnaceae)*, Vol. 4. The family of *Lemnaceae*-a monographic study. Vol. 2. Veröffentlichungen des Geobotanischem Institutes ETH, Stiftung Rübel, Zürich, Zwitzerland, 1987, 21.
2 A. Ferrer, L. Urribarrí, J. Peters, O. González, M. Terán, D. Chacón and A. Bravo, *Ciencia, LUZ*, 2006, **14**, 331.
3 L. Urribarrí, A. Ferrer and A. Colina. *Appl. Biochem. Biotechnol.*, 2005, **121/124,** 721.
4 Covenin 1156-79. *Alimentos para animales.* Determinación de Humedad. 1983.
5 A.O.A.C, *Official Methods of Analysis,* 13th, Association of Official Analytical Chemists, Maryland, 1990.
6 H. Goering and J. Van Soest, *Forage fiber analysis* (apparatus, reagents, procedures and some applications). Agricultural Handbook N° 379. Ars Usda. Washington DC, 1970.
7 A. Ferrer, F. Byers, B. Sulbarán-de-Ferrer, B. Dale and C. Aiello, *Appl. Biochem. Biotechnol.*, 2002, **98/100**, 123.
8 O. Lowry, N. Rosebrough, A. Farr and R. Randall, *Anal. Chem.*, 1965, **16**, 190.
9 L. Urribarrí, D. Chacón, O. González and A. Ferrer, *Appl. Biochem. Biotechnol.*, 2009, **153**, 94.

DIFFUSION OF BIOACTIVE PEPTIDES FROM CHITOSAN-BASED EDIBLE FILMS – EFFECTS OF TEMPERATURE AND PEPTIDES MOLECULAR WEIGHT

A.C. Pinheiro,[1] A.I. Bourbon,[1] M.A.C. Quintas,[1,2] C. Rocha,[3] J.A. Teixeira[1] and A.A. Vicente[1]

[1]IBB, Universidade do Minho, Campus de Gualtar, 4710-057 Braga, Portugal
[2]CBQF-INTERFACE A4, Universidade Católica Portuguesa, R. Dr. António Bernardino de Almeida, 4200-072 Porto, Portugal
[3]ESTG, Instituto Politécnico de Viana do Castelo, Avenida do Atlântico, Apartado 574, 4900-348 Viana do Castelo, Portugal

1 INTRODUCTION

Edible films have the potential to improve the quality and to extend the shelf life of food products, without impairing consumer acceptability. They are usually applied to control moisture transfer, limit gas transport, retard oil and lipid migration, prevent solute or flavour absorption, carry bioactive agents (such as antimicrobials and antioxidants) and improve structural integrity of foods.[1,2] Chitosan, an aminopolysaccharide obtained from chitin by *N*-deacetylation, is a promising edible film component due to its film-forming capacity and good mechanical and oxygen barrier properties. In addition, chitosan has intrinsic antimicrobial activity and inhibits the growth of a wide variety of bacteria[3] and has been proved to be nontoxic, biodegradable, and biocompatible. Chitosan films have also shown to be suitable for controlled release of drugs.[4,5]

A diverse range of biological activities can be found in many food protein sources, namely: i) the hydrolysates produced from whey protein concentrate (HWPC) show antihypertensive properties[6] or ii) glycomacropeptide (GMP), a peptide derived from k-casein, which exhibits antibacterial activity, modulates the immune systems responses and regulates the blood circulation.[7]

Diffusivity of an additive in an edible layer is proportional to how quickly the additive moves in the system and the time required to reach steady state. The diffusion mechanisms of bioactive compounds in films are critical. The diffusion of low molecular weight compounds in polymeric matrices is governed by two simultaneously occurring phenomena and previous research reported experimental observations ranging these limiting theories: i) A substantially stochastic phenomenon (related to Brownian motion), where the penetrant flows exclusively driven by a concentration gradient (usually associated with Fick's diffusion) and ii) a relaxation phenomenon driven by the distance of the local system from the equilibrium.[8]

In this study, chitosan-based antimicrobial packaging films were used as carriers for bioactive peptides (HWPC of different molecular weights and GMP) and the diffusion of these in water was studied and modelled at different temperatures (4, 16, 25 and 35 ºC). A global model was developed, predicting both temperature and molecular weight effect on peptide release.

2 METHOD AND RESULTS

2.1 Peptides release kinetics

Chitosan edible films were prepared using 2 % (w/v) chitosan and 1 % lactic acetic solution and glycerol at 1 % (w/v) and Tween 80 at 0.2 % (w/v) were added. Bioactive peptides were added to film forming solutions (10 mg/ml (dry basis)) - glycomacropeptide (GMP), and two fractions of peptides produced from whey protein concentrate hydrolysis (HWPC). Peptide molecular weight was determined using gel electrophoresis SDS-PAGE: 1 and 3.5 kDa for the HWPC peptides and 17.5 kDa for GMP.
The peptide-containing films were cut in squares (2 x 2 cm; app. weight 3.2 g) and the film thickness was measured, at six different points, using a digital micrometer.
Film squares were immersed into distilled water (1:100), at a constant temperate (4, 16, 25, 35 °C) and the peptides release was monitored until an equilibrium value was reached.
Peptides concentration was assessed by measuring the absorbance at 280 nm and 230 nm for HWPC and GMP, respectively.

2.2 Modelling diffusion in polymeric matrices

In order to evaluate the release mechanism of bioactive peptides from chitosan matrices at different temperatures, two kinetic models were investigated. Non-linear regression analysis was performed using STATISTICA 7.0. The Levenberg-Marquadt algorithm for the least squares function minimisation was used. The statistical indicators of the quality of the regression, coefficient of determination (R^2) and the mean square error (MSE, i.e. the sum of squares of residuals divided by the corresponding degrees of freedom), and the 95% Standard Error of the parameters (SE) were obtained directly from the software. Adjusted R^2 was calculated in order to account for the number of explanatory terms in of both models. The precision of the estimated parameters was also evaluated by the Standardised Half Width (SHW), which was defined as the ratio between the 95% Standard Error and the value of the estimate.
The kinetic models used were:

- Korsmeyer & Peppas model:[9]

$$\frac{M_t}{M_\infty} = k \times t^n$$

where M_t/M_∞ is the fraction of drug release at time t (first 60%), k is a constant incorporating the properties of the macromolecular polymeric system and the drug and n is an exponent used to characterize the transport mechanism.

- Flores *et al.* model:[10]

$$\frac{M_t}{M_\infty} = X_F \times \left\{-0.81 \times \exp\left[-(Const \times 10^{-10}) \times 3.28 \times 10^8 \times t\right]\right\} + (1 - X_F) \times \left[1 - \exp\left(-\frac{t}{\tau}\right)\right]$$

where *Const* is $D \times 10^{-10}$ is diffusion coefficient, τ is the relaxation time associated to polymer relaxation, X_F is a measure of the deviation of the transport mechanism from the ideal Fickian behaviour, which range from 0 (case II transport) to 1 (Fick's second law).
The results of the fitting procedure are listed in Table 1.

As can be inferred from Table 1, for both models, the parameters were estimated with adequate precision (the maximum SHW was about 60 %). F-test showed that, in most experimental conditions, the Korsmeyer & Peppas model and Flores model have similar error of prediction. Figure 1 also shows adequate description of the experimental observations, for each model, in the considered range.

Table 1 *Results of Korsmeyer & Peppas and Flores et al. models fitting to experimental data. Regression quality evaluated on the basis of MSE and adjusted R^2. Estimates precision is evaluated using SHW%. Models are compared based on F-test.*

Model	Peptide	T(°C)	Model Parameters						Regression Results		F-test p-value
			n	SWH	k	SWH			MSE	Adj R^2	
Korsmeyer & Peppas	GMP	4	0.47	16.03	2.20E-02	45.45			4.23E-03	0.8574	0.44
		16	0.39	12.34	3.09E-02	32.36			5.15E-03	0.8568	0.12
		25	0.31	9.71	4.75E-02	20.63			1.28E-03	0.9429	9E-07
		35	0.25	9.20	7.20E-02	17.50			1.75E-03	0.9383	2E-09
	HWPC 3kDa	4	0.67	10.42	1.02E-02	38.31			1.40E-03	0.9520	0.37
		16	0.65	13.02	1.28E-02	45.31			2.10E-03	0.9243	0.08
		25	0.83	7.44	6.20E-03	32.26			6.15E-04	0.9785	0.82
		35	0.69	9.81	1.50E-02	32.00			9.31E-04	0.9611	0.02
	HWPC 1kDa	4	0.53	14.85	1.75E-02	46.29			3.46E-03	0.8733	0.17
		16	0.54	16.61	1.71E-02	63.16			3.59E-03	0.8589	0.44
		25	0.56	13.21	1.55E-02	42.58			2.51E-03	0.9091	0.71
		35	0.60	10.20	1.62E-02	32.72			1.71E-03	0.9502	0.61
			Const	SWH	τ	SWH	X_F	SWH	MSE	Adj R^2	
Flores *et al.*	GMP	4	3.71E-03	19.54	491.61	12.64	0.4195	11.12	4.10E-03	0.9696	
		16	3.34E-03	16.47	564.01	19.02	0.5594	9.38	4.05E-03	0.9626	
		25	3.04E-03	16.59	400.55	26.25	0.6307	8.75	4.23E-03	0.9559	
		35	2.88E-03	14.42	219.31	33.95	0.7078	6.65	6.42E-03	0.9167	
	HWPC 3kDa	4	8.60E-03	64.97	398.26	7.64	0.1596	37.39	1.28E-03	0.9901	
		16	6.76E-03	59.44	318.75	7.98	0.1757	28.42	1.38E-03	0.9886	
		25	1.01E-02	50.94	259.94	6.15	0.1625	26.04	8.49E-04	0.9934	
		35	1.86E-02	28.53	187.73	5.30	0.2431	16.69	4.96E-04	0.9958	
	HWPC 1kDa	4	1.03E-02	40.76	508.63	14.20	0.3763	27.22	2.64E-03	0.9769	
		16	9.19E-03	52.29	462.93	15.50	0.3336	33.32	3.46E-03	0.9706	
		25	1.08E-02	42.73	481.84	15.27	0.3790	27.73	2.99E-03	0.9746	
		35	1.04E-02	32.35	300.11	12.54	0.3903	17.86	1.87E-03	0.9836	

This allowed the conclusion that both models satisfactorily fit the experimental data, suggesting that the applied models can be used to obtain useful information on the release mechanism of peptides with different molecular weight from polymeric matrix.

2.3 Global model fitting

The effects of temperature and molecular weight can be assessed by inclusion in model parameters, on the basis of experimental data and fitting procedures. This mixed-model, that includes the influence of one or more main effects and significant interactions, accounts empirically for a mathematical description of different phenomena in foods.

In this work, the effect of temperature was investigated on both models parameters. Table 2 shows the empirical relationships found. The temperature effect on models parameters was firstly determined. The parameters resulting from the subsequent application of a one-step fitting procedure allowed the expression of molecular weight influence on peptide release and, finally, the fitting of global models to the experimental data (Table 2).

Table 2 *Effect of temperature and molecular weight on parameters of Korsmeyer & Peppas and Flores models for bioactive peptides release.*

Model	One Step Fitting		Global Fitting	
	Parameter	Temperature effect	Parameter	Molecular weight effect
Korsmeyer & Peppas	k	Constant	k	a*exp(b*MW)
	n	Constant	n	c – d*MW
Flores *et al.*	Const	Constant	Const	e – f*MW
	τ	a + b*temperature	a	Constant
			b	c – d*MW
	X_F	c – d*temperature	c	Constant
			d	Constant

The results from fitting the two global models are listed in Table 3. Figure 1 shows an example of both global models description of the experimental data.

Table 3 *Results of global Korsmeyer & Peppas and Flores et al. models fitting to experimental data. Regression quality evaluated on the basis of MSE and adjusted R^2. Estimates precision is evaluated using SHW%. Models are compared based on F-test.*

Model	Parameter	Estimate	SWH (%)	Adjusted R^2	MSE	F-test p-value
Korsmeyer & Pepas	a	1,97E-02	24,1	0,7985	6,25E-03	0,0049
	b	6,44E-02	25,6			
	c	0,5455	8,0			
	d	1,50E-02	18,9			
Flores *et al.*	a	0,3645	9,2	0,9529	5,15E-03	
	h	7,40E-04	15,0			
	g	2,60E-03	77,3			
	e	1,52E-02	18,8			
	i	7,06E-04	23,2			
	c	502,80	8,5			
	d	7,4335	21,8			

As shown in Table 3 and Figure 1, the parameters were estimated with adequate precision and the observed behaviour was better predicted by Flores *et al.* model.

Regarding Flores *et al.* model parameters, the diffusion (evaluated by *Const*) was only dependent on the molecular weight, while the polymer relaxation (τ) was only dependent on temperature, thus being a material characteristic. Deviation from Fickian behaviour (X_F) was dependent on both temperature and molecular weight.

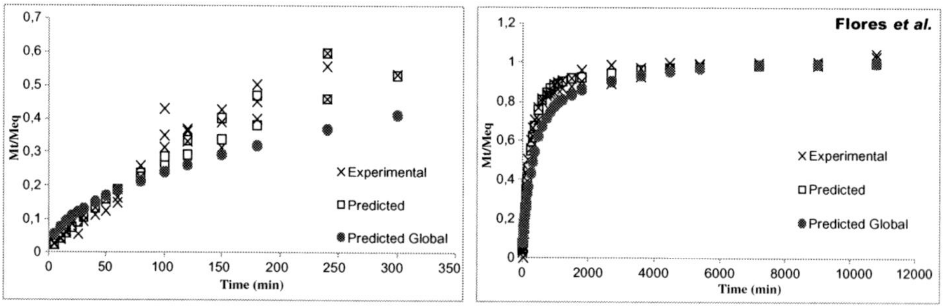

Figure 1 *Fit of Korsmeyer & Pepas and Flores et al. models to the experimental data. Example is for HWPC (molecular weight 3.5 kDa) release from a chitosan film at 25°C.*

3 CONCLUSION

The experimental data was adequately described by mathematical models, which account for both Fickian diffusion of the peptides and relaxation of polymer's matrix due to swelling. Global models were developed, predicting both temperature and molecular weight effect on peptide release.

These results will contribute to the establishment of an approach to optimize the films based on both bioactivity and diffusion properties of the bioactive compounds, aiming at a higher benefit for the product/consumer.

References

1. K. Sanjurjo, S. Flores, L. Gerschenson and R. Jagus. *Food Res. Int.,* 2006, **39**, 749.
2. S. Flores, A.S. Haedo, C. Campos and L. Gerschenson. *Eur. Food Res. Technol.*, 2007, **225**, 375.
3. A. Bégin and M-R. Van Calsteren. *Int. J. Biol. Macromol.*. 1999, **26**, 63.
4. F-L. Mi, N-L. Her, C-Y. Kuan, T-B. Wang and S-S. Shyu. *J. Appl. Polym. Sci.*, 1997, **66**, 2495.
5. Y. Pandya and D. Knorr. *Process Biochem.*, 1991, **26**, 75.
6. M.M. Mullally, H. Meisel and R.J. FitzGerald. *Int. Dairy J.*, 1997, **7**, 299.
7. C. Thomä-Worringer, J. Sorensen and R. López-Fandiño. *Int Dairy J.*, 2006, **16**, 324.
8. M.A. Del Nobile, G. Mensitieri, P.A. Netti and L. Nicolais, *Chem. Eng. Sci.*, 1994, **49**, 633.
9. R. W. Korsmeyer, N. A. Peppas. "Macromolecular and modeling aspects of swelling-controlled systems". In: *Controlled Release Delivery Systems,* eds., T.J. Roseman, S.Z. Mansdorf, Marcel Dekker, New York, 1981, p77.
10. S. Flores, A. Conte, C. Campos, L. Gerschenson and M. Del Nobile. *J. Food Eng.*, 2007, **81**, 580.

FUNCTIONAL PROPERTIES OF *GLEDITSIA TRIACANTHOS* SEEDS EXTRACTS AND THEIR INCORPORATION INTO GALACTOMANNAN FILMS FOR FOOD APPLICATIONS

M.A. Cerqueira, B.W.S. Souza, J.T. Martins, J.A. Teixeira and A.A. Vicente*

IBB – Institute for Biotechnology and Bioengineering, Centre of Biological Engineering, Universidade do Minho, Campus de Gualtar, 4710-057 Braga, Portugal.

1 INTRODUCTION

Nowadays there is an increasing interest in identifying antioxidant properties in products from natural sources to be used e.g. in food preservation or in health-enhancing foods. Edible films can provide additional protection for food, while being a fully biodegradable, environmental friendly packaging system. They can be used to improve food quality and prolong shelf-life of fresh products and they can be used to carry functional ingredients such as antioxidants, antimicrobials, nutrients, and flavours to further enhance food stability, quality, functionality and safety.[1,2] Some publications reported that seeds could be considered a valuable source of phenolics, possessing beneficial activities such as antioxidant, anti-carcinogenic, antimicrobial, anti-mutagenic and anti-inflamatory.[3]

Previous publications have already used galactomannans from commercial and non-traditional origins as a source for films and coatings production.[4,5] The present work had three objectives: to evaluate the total phenolic content and antioxidant properties of ethanolic and water extracts from *G. triacanthos* seeds; to demonstrate whether galactomannan films are suitable to incorporate antioxidant compounds for further application in the food industry; and to show how the main film properties can change with galactomannan and extract concentrations.

2 METHOD AND RESULTS

2.1 Extraction from galactomannan and extracts recovery

The polysaccharide extraction was performed as described in Cerqueira *et al.*[6] with some changes in the pre-treatment processes. In the Pre-treatment A (PTA) the seeds were removed from the pods, cleaned and placed in a blender, where they were mechanically broken. Following this operation, they were suspended in ethanol (purity 99.8%, Riedel-de Haën, Germany) in a proportion of 1:3 (seeds:ethanol) at 70 °C during 15 min. The ethanol was decanted and the endosperm was manually separated from the germ and the hull. Then distilled water was added in a proportion of 1:5 (endosperm:water) and the suspension was left to rest for approximately 24 h. The ethanol decanted in this process was called extract A.1. Then water, in a proportion of 1:10, (suspension:water) was added and mixed in a blender for 5 min. In the Pre-treatment B (PTB) the seeds were milled and the same proceeding of PTA was adopted, being the ethanol decanted in this process called extract B.1, and the water left during 24 hours removed being called B.2. Then water, in a proportion of 1:15 (suspension:water) was added and mixed in a blender for 5 min. In the Pre-treatment C the milled seeds were suspended in ethanol in a proportion of 1:3 at 70 °C during 6 hours (using a soxhlet system). The ethanol decanted in this process was called C.1, and the water left during 24 hours was removed being called C.2. Then water, in a proportion of 1:15 was added and mixed in a blender for 5 min. In the purification processes the endosperm mixed in the blender was filtered through a nylon net followed by a centrifugation step at 3800g (Sigma 4K, B. Braun, Germany) during 20 min at 20 °C. The precipitation of the galactomannan was achieved by adding the supernatant to ethanol at a ratio of 1:2. The ethanol was decanted and the precipitated galactomannan was lyophilized and kept in a dry place until further use. The ethanol decanted in this process was called extract A.3, B.3 and C.3 according to the pre-treatment used. All the extracts from the different extraction steps were filtered (Whatman, 70 mm), and concentrated in a rotary evaporator at 60 °C. The extracts were stored in the dark at 5 °C until further use.

The polysaccharide yield (*PY*) was determined at the end of the purification for an initial mass of 50 g seeds. *PY* represents the total yield of the extraction and purification processes and was calculated dividing the mass of lyophilized galactomannan (*ml*) by the initial mass of the seeds (*mi*), expressed as the % $g_{galactomannan}$ 100 g_{seeds}^{-1}. The extracts yield (*EY*) was determined at the end of each recovered process *pi* (where *p* is the pre-treatment used A, B and C, and *i* the number of the process 1 (for process A), 2 (for process B) and 3 (for process C)). EY_{pi} was calculated dividing the mass of the recovered extract (m_{re}) by the initial mass of seeds (m_i), being expressed as the % $g_{extract}$ 100 g_{seeds}^{-1}.

Table 1 *Galactomannan yield (PY), extract yield (EY), total phenolic content (TPC), radical scavenging activity (RSA) and IC_{50} of the analysed extracts; RSA and IC_{50}.*

Sample	*PY* (%)	*EY* (%)	*TPC* ($mg_{galicacid}\ g_{extract}^{-1}$)	*RSA* (%)	IC_{50} ($mg\ mL^{-1}$)
A.1	23.99	0.038	6.13 ± 0.69^{a}	18.77 ± 3.22^{c}	13.31 ± 0.67^{c}
A.3		0.074	10.79 ± 0.99^{c}	71.59 ± 3.60^{b}	1.40 ± 0.37^{d}
B.1	20.20	0.138	4.93 ± 0.42^{a}	61.88 ± 5.32^{a}	3.94 ± 0.49^{ab}
B.2		0.020	10.7 ± 0.96^{bc}	70.03 ± 4.29^{b}	3.40 ± 0.31^{a}
B.3		0.024	9.90 ± 0.81^{b}	71.11 ± 4.42^{b}	3.20 ± 0.18^{a}
C.1	16.59	0.052	5.86 ± 0.56^{a}	62.11 ± 1.98^{a}	4.17 ± 0.32^{b}
C.2		0.060	12.34 ± 1.51^{c}	70.99 ± 0.48^{b}	3.48 ± 0.13^{a}
C.3		0.066	9.82 ± 0.33^{b}	66.77 ± 5.10^{ab}	3.45 ± 0.28^{a}

[a-d] Different superscript letters in the same column indicate a statistically significant difference (Tukey test $p<0.05$).

Table 1 presents the values from polysaccharide extraction yield (*PY*) showing that the pre-treatment can influence the final yield of the galactomannan extraction. The galactomannan obtained thought PTA showed the highest value of *PY* (23.99 %), followed by PTB (20.20 %) and PTC (16.59 %). These differences may be explained by the fact that in PTA the seeds were only broken and not totally milled as in the other two pre-treatments. Further, the difference registered between the values of *PY* for PTB and PTC could be attributed to the 6 hours of extraction used in PTC, when compared with the 15 minutes used in PTB and PTA. The values of *PY* achieved with PTA are in good agreement with other results presented by Sciarini *et al.*[7] and by Cerqueira *et al.*[4] Extracts yield results are also shown in Table 1, being B.1 the extract with the highest *EY* value

(0.138%). The combination of the milling with a lower time of extraction, in the first step, provided higher values of *EY* for PTB, but in the following steps (B.2 and B.3) the values were lower when compared with PTA and PTC. A.3 has the second highest value of *EY* that might be explained by the absence of the second extraction step. In this case, the water left during 24 hours is used in the hydration of the galactomannan. The subsequent precipitation with ethanol thus includes the compounds than are not removed in step two as in the other treatments. Considering the total *EY* values, treatment B was the most efficient with a total value of 0.182%, followed by the treatments C and A with total values of 0.178% and 0.112% respectively. Not surprisingly, in those pre-treatments where the milling of the seeds is one of the steps, the total *EY* was higher than in the pre-treatment where the seeds were only broken.

2.2 Total phenolic content (*TPC*) and antioxidant activity

The determination of *TPC* in extracts and films was done by a UV spectrophotometer (Varian-UV–VIS Spectrophotometer, Germany), based on a colorimetric oxidation/reduction reaction, as described by Skerget *et al.*[8] The oxidizing reagent used was Folin–Ciocalteu reagent (Panreac, Spain). The results were expressed as g of gallic acid (Sigma, Germany). The DPPH-scavenging potential of the different extracts and films was measured, based on the scavenging ability of stable 1.1-diphenyl-2-picrylhydrazyl (DPPH) (Sigma, Germany) radicals by *G. triacanthos* extracts. The ability of extracts to scavenge DPPH radicals was determined by the method of Blois.[9] The absorbance was measured at 517 nm (Varian-UV–VIS Spectrophotometer, Germany) and the activity was expressed as percentage DPPH-scavenging activity relative to the control. The IC_{50} value was also calculated as the concentration of the compounds that caused a 50% inhibition of the RSA. All experiments were carried out in triplicate.

The extracts from the first step of the pre-treatments (A1, B1 and C1) present in all cases the lowest values of *TPC*, with statistically significant difference ($p<0.05$) when compared with *TPC* values of the others steps (Table 1), being this explained by the ethanol extraction performed in this step. This indicates that the phenolic compounds present in *G. triacanthos* extract were mostly constituted by water-soluble compounds. This fact is confirmed by the results for the extracts from step 2 (B.2 and C.2) that present the highest ($p<0.05$) value of *TPC*. In this step the seeds were left in water during 24 h. In PTA, the absence of the step 2 makes of step 3 extraction using water followed by ethanol, which explains the higher values ($p<0.05$) of *TPC* for extract A.3 when compared with A.1. The obtained values are in agreement with other studies, using other seeds species.[10,11]

The DPPH radical is widely used to evaluate the free-radical scavenging capacity of antioxidants;[12] in this work, it is presented in terms of the radical scavenging activity (*RSA*) and IC_{50}. Table 1 shows that the highest ($p<0.05$) values of *RSA* were achieved by the fractions A.3, B.2, B.3, C.2 and C.3, and the lowest value of IC_{50} was found for fraction A.3 ($p<0.05$). All the fractions corresponding to the first step of the pre-treatments (A.1, B.1 and C.1) have the lowest values ($p<0.05$) of *RSA* and, consequently, the highest values of IC_{50} confirming, in the case of fraction A.1, that the milling of the seeds before ethanol boiling in the pre-treatment process allows a more efficient antioxidant extraction thus increasing the antioxidant activity of those extracts. However, between fractions B.1 and C.1 no statistical significant differences could be observed ($p>0.05$), which leads to the conclusion that the time of extraction seems not to be an important parameter for antioxidant extraction. Fraction A.3 presents a significant antioxidant activity (and the lowest IC_{50}) when compared with the other fractions ($p<0.05$). The values of IC_{50} are also in good agreement with other types of extracts.[12,13,14,15] A relationship between the *TPC*,

RSA and IC_{50} can be observed in Table 1, where higher values of *TPC* show a positive correlation with the antioxidant activity. The DPPH scavenging activity of flavonoids and phenolic acids has been studied and confirmed in other works.[16,17] The extract used in the film formulation was selected as the correspondent to the fraction A.3 where the best values of *TPC*, *RSA* and IC_{50} (10.79 $mg_{galicacid}\ g_{extract}^{-1}$, 71.59% and 1.40 mg mL^{-1}, respectively) are obtained simultaneously.

The values of total phenolic content (*TPC*) and radical scavenging activity (*RSA*) in galactomannan films containing the extract as a function of the extract and galactomannan concentrations are shown in Table 2. As expected, these results demonstrate that galactomannan films with the incorporation of the extract present in both cases improved *TPC* and *RSA* when compared with the films without extract, due to the entrapment of the extract compounds in the film. The increase of *TPC* and *RSA* is generally proportional to the amount of extract added. To films with 0.5% of galactomannan and 0% of extract was detected a value 0.63 ± 0.19 $mg_{galicacid}/g_{film}$ to *TPC*. This value can be explained by the presence of phenolic compounds in the galactomannan polysaccharide that remain during the polysaccharide extraction process.

2.3 Films properties

The film formulations were based in a 2^2 level factorial design with a centre point. Galactomannan concentrations of 0.5, 1.0 and 1.5% (w/v) were used, together with extract concentrations of 0.0, 0.5 and 1.0% (w/v). A concentration of 0.5% (w/v) of glycerol (87%, Panreac, Spain) was used in all films formulations. The films were prepared by dissolving the lyophilized galactomannan in distilled water (20 ºC). Each mixture of galactomannan and glycerol was stirred for 2 h at room temperature (20 ºC) and the seeds extracts were added at the corresponding concentration; the obtained suspensions were homogenized with an Ultra-Turrax homogenizer (T 25, Ika-Werke). The films were prepared with a constant amount (28 mL) of solution that was cast onto a 9 cm diameter glass plate. The films were dried in an oven at 35 °C for 16 hours and maintained at 20 °C and 0% RH, until further use. The film thickness was measured with a digital micrometer (No. 293-5, Mitutoyo, Japan). The mean values were used to calculate water vapour permeability (*WVP*) and are presented in Table 2. The measurement of *WVP* was performed gravimetrically based on ASTM E96-92 method.[18] The colour of the films was determined with a Minolta colorimeter (Cr 400; Minolta, Japan). A white standard colour plate (Y=93.5, x=0.3114, y=0.3190) for the instruments' calibration was used as a background for colour measurements of the films, and the L^*, a^*, b^* values of each film were evaluated by reflectance measurements. The opacity of the samples was determined according to the Hunter lab method, as the relationship between the opacity of each sample on a black standard.

The *WVP* values ranged from 5.02 ± 0.32 x 10^{-11} (0.5% galactomannan) to 12.05 ± 1.09 x 10^{-11} $g{\cdot}m^{-1}\ s^{-1}\ Pa^{-1}$ (0.5% galactomannan and 1.0% extract), being significantly different between each other ($p<0.05$), and increasing with the increase of extract concentration (Table 2). The hydrophilic-hydrophobic ratio of the film constituents plays an important role in the water vapour transfer process.[19] Seed extracts contain phenolic acids and flavonoids that contain polar compounds, therefore contributing to improve the hydrophilic properties of the films thus increasing the *WVP*. Also the hydrophobic properties of the extracts can contribute to extend intermolecular interactions of the structural matrix in the galactomannan film, allowing the water vapour to pass through the film. Similar results were obtained by Nuthong *et al.*[20] with porcine plasma protein based films where the addition of tannic, caffeic and ferulic acid increase significantly the values of *WVP*.

The opacity means a smaller transparency, important to control the incidence of light on food. Opacity values increase with the concentration of polysaccharide for all the studied films (Table 2). Also, the addition of extract to the films leads to an increase of the opacity values, essentially due to the colour of extract added. Also Gómez-Guillén *et al.*[21] showed that the incorporation of *murta* extracts to tuna-fish gelatin films decrease the transparency of the resulting films. Table 2 shows the highest values (p<0.05) of a^* and b^* for the films containing extracts when compared with the films without extracts. In fact, the films with extract present a darker and yellower appearance as evidenced by the lower value of L^* and the higher values of b^*. The films with extract can have advantages to due their higher opacity, which can affect the light transmission through the film, having better light barrier properties which are advantageous when the light incidence is to be avoided.

Table 2 *Values of thickness, RSA, TPC, WVP, opacity, L^*, a^* and b^* of the films.*

Galactomannan % (w/v)	0.5	0.5	1.5	1.5	1.0
Extract % (w/v)	0.0	1.0	0.0	1.0	0.5
Thickness (mm)	0.045±0.001^a	0.110±0.004^b	0.052±0.002^c	0.105±0.006^b	0.089±0.012^d
RSA (%)	0 ^{a}	36.17±0.89 ^{b}	0 ^{a}	23.40±5.25 ^{c}	12.77±0.34 ^{d}
TPC (mg $_{galicacid}$ g $_{film}^{-1}$)	0.63±0.20 ^{a}	3.2±0.51 ^{b}	0 ^{c}	1.83±0.48 ^{d}	0.96±0.10 ^{e}
WVP x 10^{-11} (g·m^{-1} s^{-1} Pa^{-1})	5.02±0.32 ^{a}	12.05±1.08 ^{b}	6.54±0.37 ^{c}	9.31±0.67 ^{d}	7.86±1.16 ^{c}
Opacity (%)	4.35±0.57^a	11.02±1.05^a	4.94±0.09^a	20.09±1.68^a	16.89±0.80^a
L^*	88.62±1.29^a	69.18±3.88^a	86.77±2.11^a	68.72±0.28^a	81.42±2.30^a
a^*	5.66±0.19^a	10.73±2.15^a	5.85±0.68^a	11.73±0.15^a	4.49±1.77^a
b^*	11.49±1.56 ^{ab}	63.76±1.56 ^{ab}	14.94±2.20 ^{ab}	57.09±4.12 ^{ab}	49.59±4.01 ^{ab}

$^{a-e}$ Different superscript letters in the same row indicate a statistically significant difference (Tukey test p<0.05).

3 CONCLUSIONS

The results of this work have shown that *G. triacanthos* seeds extracts can be used as a natural source of phenolic compounds and antioxidants; it has also shown that films produced from galactomannans of *G. triacanthos* are suitable to incorporate those antioxidant compounds for further application in the food industry, while also showing how the main film properties can change with galactomannan and extract concentrations.

Acknowledgments

M.A. Cerqueira and J. T. Martins are recipient of fellowships from the Fundação para a Ciência e Tecnologia (FCT, Portugal) (SFRH/BD/23897/2005 and SFRH/BD/32566/2006, respectively) and the author B.W.S. Souza is the recipient of a fellowship from the Coordenação de Aperfeiçoamento de Pessoal de Nível Superior (CAPES, Brazil).

References

1 D.S. Lee in *Innovations in Food Packaging,* ed. J. Han, Elsevier Science & Technology Books, 2005, 108.
2 D. Lin and Y. Zhao, *CRFSFS*, 2007, **6**, 60.

3 M.A. Al-Farsi and C.Y. Lee, *Food Chem.*, 2008, **108**, 977.
4 M.A Cerqueira, A.M.P. Lima, B.W.S. Souza, J.A. Teixeira, R.A. Moreira A.A. Vicente, *J. Agric. Food Chem.*, 2009, **57**, 1456.
5 M.A. Cerqueira, A.M.P. Lima, J.A. Teixeira, R.A. Moreira and A.A. Vicente, *J. Food Eng.*, 2009, **94**, 372.
6 M.A. Cerqueira, A.C. Pinheiro, B.W.S. Souza, A.M.P. Lima, C. Riberio, C. Miranda, J.A. Teixeira, R.A. Moreira, M.A. Coimbra, M.P. Gonçalves and A.A. Vicente, *Carbohydr. Polym.*, 2009, **77**, 408.
7 L.S. Sciarini, F. Maldonado, P.D. Ribotta, G.T. Pérez and A.E. Léo, *Food Hydrocolloids*, 2009, **23**, 306.
8 M. Skerget, P. Kotnik, M. Hadolin, A. Rizner-Hras, M. Simonic and Z. Knez, *Food Chem.*, 2005, **89**, 191.
9 M.S. Blois, *Nature*, 1958, 1199.
10 S. Khokahar and S.G. Magnusdottir, *J. Agric. Food Chem.*, 2002, **50**, 3713.
11 A.A. Elzaawely, T.D. Xuan, H. Koyama and S. Tawata, *Food Chem.*, 2007, **104**, 1648.
12 J. Zhenbao, T. Fei, G. Ling, T. Guanjun and D. Xiaolin, *LWT Food Sci. Technol.*, 2007, **40**, 1072.
13 L. Majhenič, M. Škerget and Z. Knez, *Food Chem.*, 2007, **104**, 1258.
14 H.Y. Kil, E.S. Seong, B.K. Ghimire, I-M. Chung, S.S. Kwon, E.J. Goh, K. Heo, M.J. Kim, J.D. Lim, D. Lee and C.Y. Yu, *Food Chem.*, 2009, **115**, 1234.
15 B. Subhasree, R. Baskar, R.L. Keerthana, R.L. Susan and P. Rajasekaran, *Food Chem.*, 2009, **115**, 1213.
16 J.F.M. Da Silva, M.C. Souza, S.R. Matta, M.R. De Andrade and F.V.N. Vidal, *Food Chem.*, 2006, **99**, 431.
17 H. Hotta, S. Nagano, M. Ueda, Y. Tsujino, J. Koyama and T. Osakai, *BBA-General Subjects*, 2002, **1572**, 123.
18 V. Guillard, B. Broyart, C. Bonazzi, S. Guilbert and N. Gontard, *J. Food Sci.*, 2003, **68**, 2267.
19 J. Gómez-Estaca, L. Bravo, M.C. Gómez-Guillén, A. Alemán and P. Montero, *Food Chem.*, 2009, **112**, 18.
20 P. Nuthong, S. Benjakul and T. Prodpran, *Food Hydrocolloids*, 2009, **23**, 736.
21 M.C. Gómez-Guillén, M. Ihl, V. Bifani, A. Silva and P. Montero, *Food Hydrocolloids*, 2007, **21**, 1133.

AN INVESTIGATION ON THE EFFECT OF INGREDIENTS AND EXTRUSION TEMPERATURE ON PHYSICO-CHEMICAL CHARACTERISTICS OF TOMATO ENRICHED SNACKS

Z. Dehghan-Shoar,[1] A.K. Hardacre,[1] G. Meerdink[2] and C.S. Brennan[3]

[1] Institute of Food Nutrition and Human Health, Massey University, Private Bag 11222, Palmerston North, New Zealand.
[2] Holbeach Campus, University of Lincoln, Park Road, Holbeach, Lincolnshire, PE12 7PT, UK.
[3] Department of Food and Tourism Management, Hollings Faculty, Manchester Metropolitan University, Old Hall Lane, Manchester, M14 6HR, UK.

1 INTRODUCTION

Extrusion cooking is a high temperature-short time process which can be used to produce expanded snacks. These snacks generally contain high levels of starch, that is responsible for their structure and contain little else with respect of nutritional attributes. Incorporation of tomato derivatives such as tomato paste or skin, into these products may enhance their nutritional value.

Lycopene is a nutrient that has been proposed to provide health benefits including immune system modulation, free radical scavenger and anticarcinogen properties. Daily consumption of lycopene is recommended to be 6-15 mg.[1] Another nutritionally important constituent of tomato derivatives is dietary fibre. This is the edible part of plants which resists digestion and absorption. Beneficial physiological effects such as laxation and modulation of blood glucose have been related to consumption of dietary fibre.[2]

Due to presence of fibre and sometimes sugar in tomato derivatives, extrusion at high temperature and shear conditions is needed to produce extruded products with acceptable physico-chemical characteristics;[3] but lycopene is heat labile and degrades in these conditions.[4] By changing the chemical composition of the melt using different ingredients, physical properties of the extruded products can be improved but how this affects lycopene is not known.[5]

The aim of this study was to determine the effect of extrusion processing temperature and different raw ingredients on retention of lycopene in snack foods extruded at 140 or 180°C. Lycopene content, expansion, density, moisture content and colour parameters of products manufactured from rice, wheat or corn as the starch source enriched with freeze dried tomato paste or skin powder used as the lycopene sources were investigated.

2 METHODS AND RESULTS

2.1 Methods

2.1.1 Chemical composition. Proximate analysis of ingredients was determined according to the AOAC method.[6] All measurements are reported as a mean of triplicate analysis.

Lycopene was extracted from the extruded products using hexane: acetone: ethanol (2:1:1 v/v) and the absorbance was determined at 472 nm using spectrophotometer (Shimadzu Co., Ltd., Japan). Standards were made using pure lycopene purchased from the Sigma Chemical Company (Sydney, Australia) and were used for the preparation of a calibration curve. All measurements are reported as mean of duplicate analysis.[7,8]

2.1.2 Extrusion process. Tomato paste and skin were supplied by Heinz Watties (Heinz-Watties NZ Ltd. King St. Hastings, NZ). They were freeze dried, then ground to a mesh size of 20 and blended with Sunrise rice flour, Champion brand wheat semolina (Davis Trading, Palmerston North, New Zealand) or 320 spec corn grits (Corson Grain Ltd, Gladstone Rd, Gisborne, NZ) at a level of 20%. All extruded products were made using a Clextral BC21 twin-screw laboratory scale extruder (Clextral, Firminy Cedex, France) with die temperature set at 110 or 140°C. Feed rate, water feed rate and screw speed were kept constant at 11.52 Kg/hr, 0.5 l/hr and 350 rpm. Extrusion parameters are shown in Table 1.

2.1.3 Physical characteristics. Expansion was determined by the ratio of radial diameter of the products to the die diameter and product density by the poppy seed displacement method. Colour parameters were taken at five separate points of milled samples using Minolta Chroma meter (CR200, Minolta, Japan).[3]

2.1.4 Statistical analysis. A completely randomized design was used to evaluate the results and Analysis of Variance (ANOVA) was carried out to compare the mean values. All significant differences are reported at a probability level of 0.05. All statistics were calculated using SAS software version 9.1 (SAS Inc., Chicago, IL, USA).

2.2 Results

2.2.1 Proximate composition. The chemical composition of the raw material is shown in Table 2. Wheat had close to 10% less starch compared to rice and corn. The protein content of wheat semolina was 10.7% and about 40% greater than that of rice and corn. The fat content for rice flour was about 6% of that of corn grits or wheat semolina. Due to the higher levels of protein and fat in the wheat semolina, it would be expected that the products made from this material would be denser and harder.[10]

2.2.2 Extrusion parameters. Die temperature, thrust pressure, torque and SME are shown in Table 1. The incorporation of tomato derivative reduced torque and SME values significantly due to the lubricating effect of sugar and fibre present in tomato derivatives[9] (Table 1).

2.2.3 Lycopene content. Although lycopene content in products containing tomato paste was the highest, lycopene retention in products containing tomato skin was higher (Figure 1). This is because lycopene in tomato paste is more accessible therefore degrades readily with processing. Lycopene content of wheat products containing tomato paste were significantly lower than corn and rice but this effect was not observed in products containing tomato skin (Figure 1 and Table 3). The temperature range used did not have a significant effect on lycopene content. This is similar to findings of researchers for other heat labile components.[11] SME values correlated negatively with lycopene content (r= 0.701) (Table 3).

Table 1 *Extrusion parameters for tomato enriched extruded snacks.*

Die Temperature (°C)	Starch source	Lycopene source	Thrust pressure (bar)	Torque (N.m)	SME (KJ/Kg)
110	Corn Grits	Paste	135	4.75	111
		Skin	112	3.60	106
		None	150	5.60	170
	Rice Flour	Paste	49	4.90	149
		Skin	49	4.80	146
		None	78	7.80	237
	Wheat Semolina	Paste	80	3.70	112
		Skin	149	5.30	161
		None	205	6.80	207
140	Corn Grits	Paste	27	3.70	112
		Skin	26	3.10	94
		None	55	5.00	176
	Rice Flour	Paste	45	5.00	152
		Skin	43	3.50	106
		None	78	5.90	179
	Wheat Semolina	Paste	49	3.90	118
		Skin	120	4.60	140
		None	140	5.50	167

Table 2 *Chemical composition of ingredients.*

Raw material	Moisture (%)	Fat (%)	Protein (%)	Dietary Fibre (%)	Lycopene (ppm db)	Ash (%)	Starch (%)
Rice flour	12.11 ^{a}	0.14 ^{d}	6.75 ^{d}	0.8 ^{d}	0 ^{c}	0.28 ^{c}	79.78 ^{a}
Corn grits	11.77 ^{a}	2.33 ^{b}	6.74 ^{d}	1.1 ^{d}	0.2 ^{c}	0.57 ^{c}	78.31 ^{a}
Wheat Semolina	12.48 ^{a}	2.15 ^{b}	10.72 ^{b}	2.5 ^{c}	0 ^{c}	0.56 ^{c}	69.67 ^{b}
Tomato skin	4.09 ^{b}	1.17 ^{c}	8.05 ^{c}	78.5 ^{a}	18.3 ^{b}	2.0 ^{b}	6.69 ^{c}
Tomato paste	5.96 ^{c}	2.96 ^{a}	14.14 ^{a}	19.64 ^{b}	664 ^{a}	6.74 ^{a}	6.32 ^{c}

2.2.4 Physical characteristics. Expansion of products ranged between 186 – 360%. The incorporation of tomato paste or skin decreased expansion up to 25% compared to the controls. Rice flour produced the most expanded products due to the higher starch content (Table 2 & 3).

Product density varied between 172 – 751 kg/m^3. Weights of 1 litre of products containing tomato paste were 2-fold higher than the other samples. Similar to findings of other researchers,[3,12] increase in temperature, decreased product density (Table 3).

Among colour parameters, "a" value is a useful tool to compare lycopene content of the extruded products.[3] Products containing tomato paste had higher "a" values. Increases in temperature affected "a" value negatively (Table 3).

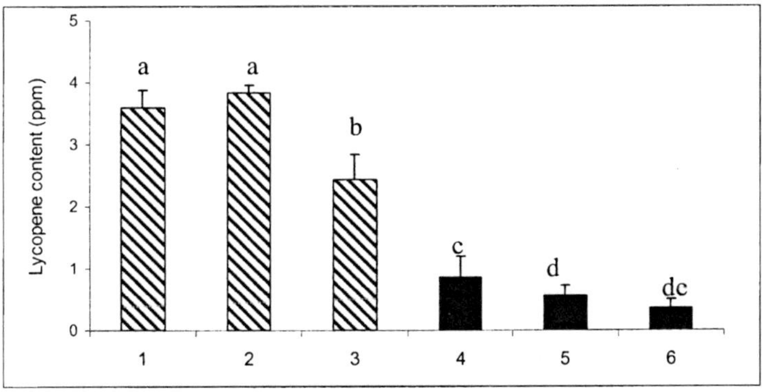

Figure 1 *Effect of ingredients on lycopene content of extruded products. 1= Corn grits and tomato paste, 2= Rice flour and tomato paste, 3= Wheat semolina and tomato paste, 4= Corn grits and tomato skin, 5= Rice flour and tomato skin, 6= Wheat semolina and tomato skin. Values with different letters are significantly different (P <0.05)*

Table 3 *Effect of treatments on physico-chemical characteristics of the extruded products.*

	SME (KJ/Kg)	Lycopene content (ppm)	Expansion (%)	Density (Kg/M3)	"a" value	Moisture content (%)
Effect of lycopene source						
Paste	132.67 b	3.33 a	242.28 b	455.61 a	9.42 b	10.58 a
Skin	114.14 b	0.59 b	205.20 c	281.78 b	6.47 ab	9.28 b
non	208.29 a	0.07 c	330.27 a	250.39 b	5.60 a	9.55 b
Effect of starch source						
Rice	175.08 a	ns*	257.41 a	ns	ns	9.15 b
Wheat	174.19 a	ns	166.57 b	ns	ns	10.47 a
Corn	134.92 b	ns	230.73 ab	ns	ns	10.29 a
Effect of temperature						
140	154.67 a	ns	ns	419.49 a	8.87 a	ns
180	138.24 b	ns	ns	313.88 b	7.89 b	ns

* ns= The effect was not significant

3 CONCLUSION

The enrichment of the extruded products with tomato derivatives can enhance their nutritional properties but may deteriorate their physico-chemical characteristics. However, careful selection of ingredients improves these characteristics and enhances the stability of nutrients such as lycopene.

References

1 Y. Kun, U.S. Lule and D. Xiao-Lin, *Food Rev. Int.*, 2006, **22**, 309.
2 C.S. Brennan, *Mol. Nutr. Food Res.*, 2005, **49**, 560.

3 A. Altan, K.L. McCarthy and M. Maskan, *J. Food Eng.*, 2008, **84**, 231.
4 M.T. Lee and B.H. Chen, *Food Chem.*, 2002, **78**, 425.
5 M.E. Camire in *Process-Induced Chemical Changes in Food.* Vol. 434, Plenum Press Div Plenum Publishing Corp. New York, 1998, pp. 109.
6 AOAC, 4th revision, 16th ed. Arlington, Va: Assoc of Official Analytical Chemists Intl. 1998.
7 D. Kaur, A.A. Wani, D.P.S. Oberoi and D.S. Sogi, *Food Chem.*, 2008, **108**, 711.
8 G. Sadler, J. Davis and D. Dezman, *J. Food Sci.*, 1990, **55**, 1460.
9 M.N. Riaz, in *Handbook of Food Science, Technology and Engineering*, ed. Y.H. Hui, CRC Press, Boca Raton, Florida, 2006, p. 4.
10 R. Guy in *Extrusion Cooking, Technologies and Application,* Woodhead Publishing, Florida, USA, 2001.
11 N. Athar, A. Hardacre, G. Taylor, S. Clark, R. Harding and J. McLaughlin, *J. Food Compos. Anal.*, 2004, **19**, 379.
12 S. Ilo and E. Berghofer, *J. Food Eng.*, 1999, **39**, 73.

Subject Index

Breinigsville, PA USA
23 May 2010
238516BV00006B/11/P

9 781847 557506